The Lihir Destiny

Cultural Responses to Mining in Melanesia

Asia-Pacific Environment Monograph 5

The Lihir Destiny

Cultural Responses to Mining in Melanesia

Nicholas A. Bainton

ANU

THE AUSTRALIAN NATIONAL UNIVERSITY

E PRESS

ANU E PRESS

Published by ANU E Press
The Australian National University
Canberra ACT 0200, Australia
Email: anuepress@anu.edu.au
This title is also available online at: http://epress.anu.edu.au/lihir_destiny_citation.html

National Library of Australia
Cataloguing-in-Publication entry

Author:	Bainton, Nicholas A.
Title:	The Lihir destiny [electronic resource] : cultural responses to mining in Melanesia / Nicholas A. Bainton.
ISBN:	9781921666841 (pbk.) 9781921666858 (eBook)
Series:	Asia-pacific environment monographs ; 5.
Notes:	Includes bibliographical references.
Subjects:	Lihirians--Social life and customs. Mineral industries--Papua New Guinea--Lihir Island--Social aspects. Lihir Island (Papua New Guinea)--Social life and customs.

Dewey Number: 995.805

Cover design and layout by ANU E Press

Cover image: Francis Dalawit addressing the crowd during the *Roriahat* feast in Kunaie village, 2009. Photograph courtesy of David Haigh.

Contents

List of Plates

List of Tables

List of Figures

Foreword

Mining communities are the subject of a rich tradition of ethnographic study. As the major industry and employer in any region where they are located, mining operations provide a physical, social and economic focal point for the anthropologist. Approaches to the subject have varied greatly. From June Nash's (1993) study of a Bolivian mining community, *We Eat the Mines and the Mines Eat Us,* to Michael Taussig's polemical and literary reflections on capitalism, greed and exploitation in *The Devil and Commodity Fetishism* (1980) and *My Cocaine Museum* (2004), there has been close scrutiny of the complex inter-relationships between mines, their owners (whether individuals or corporations), and the people who live around them or work in them.

The search for mineral wealth has long been associated with colonisation, economic exploitation and economic transformation. In contemporary Papua New Guinea, large-scale mining has become the most significant export industry, generating income for government and for the people whose lands are affected. Extractive industry is now viewed as the main means of economic development. Controversies over the environmental degradation and social disruptions generated by mining operations have hardly dampened the enthusiasm for mining. International companies continue to explore and take out leases, the government continues to facilitate their activities, and in most places local people welcome the associated prospect of 'development'. Lihir has been no exception.

My own association with Lihirians and the Lihir gold mine began just before the construction phase, in 1995. At that time the excitement was almost palpable, as people contemplated the wealth that they were sure would be generated by the project. Over the following nine years I worked as a consultant, monitoring the social changes that occurred and recording the local responses to environmental change and degradation. As I observed many of the dramatic confrontations and the innovative strategies that Lihirians adopted in their dealings with the mining company, I often wished that I could find a graduate student who would be able to engage with these changes in the sort of concerted, day-to-day, ethnographic research that is characteristic of our discipline. Nick Bainton became that ethnographer.

Participant observation has had a bad press over the last two decades – decried as oxymoronic, partial and ideologically suspect – but it has survived this intellectual buffeting. This book is testament to its continued strength as a methodology, and to the ways that direct observation of events, conversations with a variety of people, and reflection upon change over time enriches interpretative endeavours. The author's participation — in the training for 'Personal Viability', in feast preparations, and in the everyday lives of the villagers with whom he lived — generates insights and descriptions that are unavailable to a casual observer.

The book is a revised version of a doctoral thesis. It incorporates archival and historical research, and engages with debates about the ways that contemporary Melanesians construct models of their identity and culture as they embrace modernity. It tells a story of the complicated making of the 'roads' that Lihirians have taken as they strive to reach their 'destiny'. It also reveals the tensions generated, within the local community and between Lihirians and others, as they struggle to gain control over the processes of change and the wealth generated by the mine.

The social and economic changes ushered in by the mining project on Lihir have been profound. They are readily observable. When I first arrived, I was struck by the relative poverty of people there. Many children still went naked, women's clothing was usually a drab length of cloth, houses were invariably made of bush materials, and there was a meagre strip of dirt road linking a few villages. The airstrip was tiny and involved the rather tricky piloting manoeuvre of landing a small plane on an uphill slope. Now there is an airport that regularly ferries hundreds of workers to and from the island; children wear shorts and sneakers, or frilly dresses from the large Filipino-owned supermarket. Where there was an overgrown and abandoned plantation there is now a township. A well-equipped modern hospital serves the local community as well as the mining company's employees.

But as Nick Bainton demonstrates in this study, the material changes have brought with them new distinctions, new inequalities, and conflicts that were previously absent. The abundance of introduced goods, the enthusiastic embrace of modernity, and associated power struggles do not mean that customs have been abandoned or that Lihirians have 'lost' their cultural traditions, their sense of their uniqueness, or their dreams of the future. Custom, like everything else on Lihir, has been transformed. This book documents and analyses the complex interactions between local people, migrants, foreign workers and mine managers, and the ways that new values associated with a monetary economy are established. It stands as a fine contribution to the anthropology of mining communities.

Martha Macintyre
The University of Melbourne
August 2010

References

Nash, J., 1993. *We Eat the Mines and the Mines Eat Us*. New York: Columbia University Press.

Taussig, M., 1980. *The Devil and Commodity Fetishism*. Chapel Hill: University of North Carolina Press.

———, 2004. *My Cocaine Museum*. Chicago: University of Chicago Press.

Acknowledgements

This book is primarily about Lihirian responses to large-scale resource development. In the process of writing this book, from its original genesis through to its current form, I have been engaged with Lihirian lives and resource development in several different ways. As such the list of people who have helped me along the way is that much longer.

This book originated as my doctoral thesis at the University of Melbourne. My first 18 months of research in Lihir was made possible through financial support from the School of Anthropology, Geography and Environmental Studies. Thanks to the National Research Institute for organising visas and research permits in Papua New Guinea. Throughout my candidature, Monica Minnegal, Peter Dwyer and Mary Patterson provided academic support, constantly challenging me to further develop my ideas and to ask more questions. I owe a great intellectual debt to Martha Macintyre who supervised my doctoral research. Her work in Lihir opened many opportunities for me and has strongly influenced my belief in the need for a genuinely engaged anthropology. Over the years Martha has generously shared her ideas, and provided continuing encouragement and friendship.

In many ways the process of revisiting, rethinking and rewriting my earlier research has followed a less than conventional route. I have shifted from village based anthropology to gradually working more closely with the mining company in Lihir. At the same time this shifting engagement has created opportunities for closer involvement with many Lihirians. In 2007, I commenced a research fellowship at the Centre for Social Responsibility in Mining (CSRM), in the Sustainable Minerals Institute at the University of Queensland as part of a three year research partnership between the centre and Lihir Gold Limited (LGL). This position took me back to Lihir on a monthly basis as I worked with company personnel and community members on social impact studies and cultural heritage management. During my time at CSRM David Brereton was exceptionally supportive (despite all of his jokes about anthropologists!), and helped to cultivate a greater appreciation of the ways to meaningfully engage companies in social research. I am also thankful to the Sustainable Minerals Institute for financial assistance towards the completion of this book, and for a very generous Early Career Research Grant to commence research on the sacred geography of Lihir. Some of the early findings of this work have contributed towards the ethnographic descriptions presented here.

Over the past three years, the staff members of the LGL Community Liaison office have been particularly helpful. Thanks to the administration staff for organising my constant flights on and off the island, and the research staff in the social impact monitoring section for providing access to data. Luke Kabariu and Patrick Turuan who work in the LGL Cultural Information office have also provided great assistance, patiently answering my endless questions, introducing me to people and helping to facilitate research around the different islands. At various times, Elly Sawa and Walter Pondrelei have assisted with data collection. David Haigh from the LGL Media Resource Production and Training section provided excellent audio-visual support, especially for the recording of ceremonial events. I have benefited immensely from continuing conversations about Lihir and Melanesia more generally with Glenn Banks, Simon Foale, Susan Hemer, John Cook, Tim Grice, Ken Costigan, John Vail, John Burton, Anthony Regan and John Cox. I am especially thankful to Chris Ballard and Kirsty Gillespie with whom I have worked for the past few years on the development of cultural heritage management strategies in Lihir. Both have widened my knowledge of Melanesia and cultivated new research interests in Lihir. I particularly want to thank Andrew Holding who developed the partnership between CSRM and LGL which helped with the continuation of my research.

Several sections of this book have been published elsewhere. Chapter 3 originally appeared in the *Journal of Pacific History* under the title, 'The Genesis and the Escalation of Desire and Antipathy in the Lihir Islands, Papua New Guinea'. Sections of Chapter 4 appeared in *The Australian Journal of Anthropology* in an article titled, 'Men of *Kastom* and the Customs of Men: Status, Legitimacy and Persistent Values in Lihir'. Sections of Chapter 6 appeared in *Oceania* in an article titled, 'Keeping the Network out of View: Mining, Distinctions and Exclusion in Melanesia', and also in a *State Society and Governance in Melanesia* discussion paper titled, 'Parallel States, Parallel Economies: Legitimacy and Prosperity in Papua New Guinea'. I am particularly thankful to the editors of these journals for permission to use this material here. Mike Fabinyi and Deana Kemp read revised chapters and provided valuable comments and appraisal. Dan Jorgensen read the original thesis and his excellent critical commentary has helped shape the current work. I am thankful for the astute comments and suggestions provided by the two anonymous reviewers of this manuscript and for the excellent editorial work by Mary Walta. Colin Filer also read the original thesis and has been instrumental in the process of writing this book. I have benefited tremendously from his unparalleled insights and knowledge of resource development in Melanesia.

My greatest debts are to the many Lihirians who have provided hospitality, shared their stories and histories, excused my mistakes, and helped me to understand their lives. While there are too many to name, I must mention Demas and Maria, Francis and Wokpul, Benjamin Rukam, Alphonse Ama, Martin Asu, Peter and Rose Toelinkanut, the late Ludwig Lel and the late John Zipzip, Jacinta Gagala, Bah Arom, Maria Tamon, Joanna Kokomalis, Clement Papte, Jaochim Malele, Joseph Kondiak, Mathew Tongia, Laurence Klumga, Martin Bangel, the members of the LSDP Committee for sharing their views, the 2004 students at Lakuplien and Kinami primary schools, and the members of the Lihir Cultural Heritage Association. *Yel apet siwa*. Finally, my deepest thanks go to my parents for their encouragement, and to Veronica who inspires me and continually spurs me on.

Selected Tok Pisin glossary

bikman	big man: leader
bisnis	business
boi	boy: derogatory colonial term used for natives
buai	betelnut
driman	dream
grasruts benk	grassroots bank
grasruts yuniversiti	grassroots university
hamamas	happiness
hausboi	men's house
kagoboi	cargo boy, labourer
kalsa	culture
kastom	custom
kiap	patrol officer
kompensesn	compensation
las kantri	last country
laplap	sarong worn by men and women
luluai (ENB)	village policeman appointed by the colonial administration
masta	master: mostly refers to European status vis-a-vis Papua New Guineans
malanggan (NIP)	carving produced in the northern part of New Ireland Province for mortuary ceremonies
masalai	spiritual beings that inhabit the landscape
mis	shell money produced in New Ireland Province
pasin bilong Lihir	the ways of Lihir
pasin bilong tumbuna	ancestral ways
raskol	petty criminal
rot	road
siti	city
tambu	in-law
tultul (ENB)	village police assistant appointed by the colonial administration
wantok	literally 'one talk': people from the same ethnic or language group
winmoni	profit or windfall

Selected Lir glossary

aginas	happiness
balo	traditional round roofed men's house
balun kale	pig consumed during the *rarhum* feast which marks the fire roasted vegetables
a bertman	a relationship term used between fathers and children
a berpelkan	a relationship term used between cross-cousins
a berturuan	a relationship term used between friends
balun peketal	pig consumed during the *rarhum* feast which marks the decoration on the deceased
bior	linage: a group descended from a proposed female ancestor
bual	pig
bual tom	sacred pigs consumed during *rarhum* feast
dal wan pour	matrilineal blood line
erkuet	widow strangulation
hurkarat	the host(s) of a feast
ihot	stony place: refers to smaller islands in the Lihir group
ikineitz	a feast performed by children to honour their parents
iol nizenis	pigs killed during the *katkatop* feast
kanut	the corpse or spirit of the deceased
karat	the feasting cycle: also signifies the final mortuary ritual in the feasting sequence
karemiel	pigs killed during the *katkatop* feast
katkatop	initial mortuary feast held soon after the death of a person (also known as *pkepke*)
katmatanarihri	pigs killed during the *katkatop* feast
kienkien	feasting stage that marks the preparatory work in the *katkatop* and *tutunkanut* feasts
konkonla	ritual canoe used in mortuary ceremonies on the outer islands of the Lihir group
lam	broad or main
a le	shell money
liling wehien	maternal nieces
malal	cleared centre of a hamlet

malkok	small
matanlaklak	Y shaped stile for entering the men's house enclosure
mok	taboo placed on a section of beach or reef after the *rarhum* feast
mormor	traditional pole displaying the skull of a deceased clan member previously used during the final mortuary feasts
motung	senior maternal uncles
a ninambal	dream
nunglik	maternal nephew
a peketon	the image of waves crashing on the shore: used as a metaphorical concept of change arriving in Lihir
pasuki	lining up the pigs in front of the men's house during major mortuary feasts
piar	woven basket
pinari wasier	to provide hospitality for guests
Pindik	traditional secret society
pkepke	initial mortuary feast held soon after a person dies (also known as *katkatop*)
poas	affine, in-law
polpol	shelter outside of the men's house enclosure primarily used by women for cooking, socialising and sleeping
puatpes	pigs contributed by guests at a feast which also mark women's contributions to feasting
pukia	a paramount leader
rarhum	sacred feast held to commemorate significant clan members as they approach senior age
rangen	mortuary ritual held soon after a person dies (also the term for mortuary songs)
rapar	bride wealth exchange
ravomatanabual	ritual payment of pigs with shell money during mortuary feasts
rihri	men's house
Rondende	afterlife
roriabalo	ritual performance on the roof of a *balo* (round roofed men's house) during the final mortuary feast
roriahat	ritual performance on a constructed stage during the final mortuary feast
saksak	return of *mis* previously received

sio	respect
tandal	spiritual beings that inhabit the landscape
Tanori	traditional secret fishing society
tele	help given to another that will be reciprocated later
tinanakarat	a large pig presented during *katkatop* and *tutunkanut* feasts
tohie	senior male leader
Triu	traditional secret society
tolup	traditional ritual seclusion process to prepare young women for marriage
trias	pig jaws
tsiretumbawin	sub-clan
tumbawin-lam	'big people cluster' (refers to a moiety)
tumbawin-malkok	'small people cluster' (refers to a moiety)
tutunkanut	final mortuary feast for finishing obligations to the deceased
wasier	guest at a feast or visitor
wehientohie	female leader
welot	stone fence around the men's house enclosure
yebhi	'putting out the fire' (metaphorical term for cancelling debt
yel apet siwa	thankyou to all of you
zik	child, boy, son
ziktohie	first-born son recognised as a leader

Abbreviations

DEC	Department of Environment and Conservation
EFIC	Export Finance Insurance Corporation
IBP	Integrated Benefits Package
KLK	Kastom Lidas Komiti
LGL	Lihir Gold Limited
LGPMKL	Lihir Grasruts Pawa Mekim Kamap (Developing the power of Lihirian grassroots people)
LJNC	Lihir Joint Negotiating Committee
LMALA	Lihir Mining Area Landowners Association
LMC	Lihir Management Company
LSDP	Lihir Sustainable Development Plan
NDA	Nimamar Development Authority
NLGC	Namatanai Local Government Council
NRLLG	Nimamar Rural Local-Level Government
PNG	Papua New Guinea
PV	Personal Viability
SML	Special Mining Lease
TFA	Tutorme Farmers Association
TIA	Tutuvul Isakul Aisok
TKA	Tuk Kuvul Association
VDS	Village Development Scheme
VPS	Village Population System

Currency Conversion Rates

Year	1 USD equivalent in PNGK
1995	0.8
2000	2.4
2003	3.5
2007	2.9
2009	2.6

1. Introduction: New Lives for Old

> All through history men have greeted periods of change, the challenge coming from foreign ideas or other peoples, in many ways — by imitation, incorporation, rejection, transformation, trance, manifesto, retreat (Margaret Mead 1956: 16).

When a large gold mine was constructed on the main island of the Lihir group in 1995, Lihirians began to envision that the sort of life they had long dreamed of was closer to being in their grasp. For the past century, Lihirians have been beguiled by the sort of development that would genuinely enhance their lives. Unlike so many of their Melanesian compatriots, Lihirians were going to witness the realisation of past prophesies for material and social change that arose during previous social movements — albeit in decidedly unexpected ways. This large-scale resource development project has precipitated tremendous economic change and social upheaval: literally in a matter of years, Lihirians were propelled from a subsistence existence supported by sporadic copra sales into an industrialised economy. They shifted from moving on the periphery of the capitalist system to embracing one of its core activities. Given the changes that have occurred since construction of the mine, it is tempting to say that Lihirians have been completely bowled over by the juggernaut of modernity that has swept through their villages and men's houses. Yet for some time, Lihirians have been lining up ready to climb aboard and ride in the direction of the nearest trade store, car dealer, or market place. Lihirians have never eschewed the accoutrements of modernity, but have embraced different means for their acquisition to develop their lives and their culture on their own terms.

When I first began fieldwork in Kinami village on the main island of the Lihir group in mid-2003, I found myself quickly absorbed in the politics of development and cultural change. Men and women talked to me about what *was* occurring, which was frequently contrasted with what they thought *ought* to occur. People were deeply engaged in a conspicuous commentary on the state of change. After a year of sitting around smoky men's houses listening to, and engaging with, a discordance of ideas about the past, present and future, I approached the local community schools to ask the students for their impressions. I asked the students from Grades Four, Five and Six to either draw a portrait of themselves in the future, how they imagined they might look or what they might be doing, or to write a short story on what they thought the future of Lihir would be like. Three stories stood out:[1]

---

1   I have not used pseudonyms in this book. I have presented these stories in their original English; some of the other stories were written in a mixture of Tok Pisin and English. No stories were written in Lihirian.

Story 1: This story I heard from some grand parents they said long ago in the past when missionaries first arrived here on the island, they form a little group that was called Nimamar. This group was interested in talking about Lihir Island.

They said that in the future Lihir Island is going to develop. And they said, this thing like cutting copra, cocoa and so on is going to stop. People of Lihir Island will not be sweating when searching for money. They will just sit and have free money coming in. They also said that Lihir Island is going to be like a city.

To conclude, now you can see that Lihir Island is changing and all things that they said is now coming true. So you can see that people of Lihir Island will be happy in their live time (Antonia Zanates, Grade Four, Kinami Community School, 2004).

Story 2: I think people in Lihir will not be able to do customs anymore, because now when people make custom they need one thing and that is money. In the future when mining ends, I think there will be no money or many kinds of good things. Where will the people in Lihir get their money from when they want to make custom? People will not be able to travel to other places to look for pigs to make custom. That is why I really think that custom will end too (Scholastica Lendai, Grade Six, Lakuplien Community School, 2004).

Story 3: I think in the future Lihir Island will not be the same as today. There is going to be changes on our Island. In the future Lihir is going to be an enormous city. Lihirians will be living in high quality permanent buildings with facilities like stoves, TV set, video and fridges. The whole island will be using electricity instead of firewood and lamps. Their ways of earning money will change from what is today. They will no longer use things like fire wood or traditional ways of cooking.

Ways of transporting goods and people from place to place will be much easier. Each family will have a car of its own. Lihir is going to be full of *raskols* [petty criminals]. I think in the future many young men and old men will leave their *hausboi* [men's house] and live only in their high quality permanent buildings. They will find life easier and forget their traditional ways of life. Ways of education will be much different from

---

Although Papua New Guinean students are educated for the first three years in the vernacular, English is the official language used for the remainder of their education. In many cases, the national lingua franca Tok Pisin is more commonly used and more readily grasped. Tok Pisin and Lihirian terms are italicised and listed in separate glossaries.

now. Children will be learning and using computers and using modern technology. So Lihir will be different from what it is today (Francisca Bek, Grade Six, Kinami Community School, 2004).

These stories capture the central themes in this book, particularly Lihirian aspirations for development and the dependency of customary practices upon the capitalist economy. They remarkably delineate those aspects of Lihirian history, desire, and entanglement with global processes, which consume the thoughts, energies and concerns of so many Lihirians that I have come to know.

Plate 1-1: Lihirian children on the shores of change. Ladolam, looking north towards the Ailaya in Luise Harbour, circa 1983.

Photograph courtesy Rudolph Kiakpe.

Lihirian children have been exposed to the material world of modernity since their early years, and this has profoundly shaped their aspirations, dreams, fantasies, and sense of what is possible. Many of their stories and portraits depicted futures characterised by technological ease and the ability to consume and use mod-cons of the latest type, in a life typified by the trappings of modernity. They imagined themselves living in a world of choice, where the future held infinite possibilities. More importantly, they saw themselves as being different from their parents and ancestors. Some imagined themselves as rock stars, carpenters and accountants, or using computers, driving trucks in the mining pit, operating the massive shovel tractors, or owning modern homes. Girls drew themselves as air hostesses travelling regularly to Australia or as

3

Miss PNG in a beauty pageant, while boys pictured themselves in the army or as police officers chasing *ol raskol* through town. Like their older relatives, they imagined that the mine would provide the means for achieving an imagined future: it would be new lives for old.

However, their stories also reveal that the transformations taking place in Lihir are complex, and are not simply an exchange of the past for the future. In some cases, the future was less favourably described through tales of modern dystopia, with images of environmental change, social disintegration, abandoned traditions and familial breakdown. These symptoms of modern life also included famine, disease, chaos, moral bankruptcy, migration, sad memories of relocation, and the possibility that, when the mine closes, life might not actually be any better. Here the past was compared with the present, and was conspicuously apparent in an anticipated future characterised by a regression towards backwardness and traditional daily toil.[2] Their stories and portraits captured the diversity of Lihirian desires and perceptions of the past, present and future. Crucially, the differences between the stories and images from different schools expressed the ways in which the unequal distribution of wealth shapes different hopes, the intimate connection between their lives and the global economy, and their increasing dependency upon mining in order to remain active in the game of 'self-improvement'.

Talk about the future is not just a current phenomenon; Lihirians were discussing it well before mining began. Different social movements, such as the Nimamar Association described by Antonia Zanates, which had its roots in the colonial period, looked to a new utopian world order. Not everyone has been convinced that mining is the fulfilment of these prophesies, nor is there agreement about future prospects. The changes and inequalities experienced through mining have only strengthened local resolve to achieve these earlier dreams. However, Lihirians have been presented with different, complex and competing roads to their ideal land of modernisation. As these stories reveal, not only is there difference of opinion on how to reach this desired state, but there is a range of expectations (and fears) about what this new life will entail.

---

2 Sigrid Awart, an Austrian ethno-psychologist who undertook field research in Lihir in 1990–91, made use of similar material, asking school children to write a short essay on the topic: 'What would I like to do after I finish Grade Six?'. Her Masters thesis was written in German (in 1993), and she has published one article in English (in 1996) where she makes reference to these stories. The occupational desires are generally similar to those listed in the stories I collected. However, as a result of contemporary influences, the students I spoke with listed a greater range of occupations. While she notes that boys were generally more interested than girls in obtaining technical occupations off the island, I found a more even spread between the sexes, with both girls and boys generally wanting to pursue a lifestyle outside of the village. However, the most noticeable shift was the number of students seeking a subsistence existence. Awart notes that out of 101 students, 22 per cent of the girls and 16 per cent of the boys wanted to become gardeners, whereas I found only three out of 120 students indicated any desire for this type of lifestyle. I found that the students were more concerned to escape this type of existence, or else they tended to think that mine closure would signal a return to subsistence living, which was not viewed positively.

With this in mind, I have cast my gaze on the ways that Lihirians have experienced modernity, both through colonialism and within the context of large-scale resource development, on the articulation of local social, economic and political change, and on their attempts to direct their lives towards different and new futures. This book seeks to understand what historically and culturally located Melanesians desire and seek. This is the story — indeed, the history — of what happens when the cargo actually arrives.

Understanding Local Responses to Global Processes

There is now enough evidence to see that different communities (or cultural groups) have responded to large-scale industrial development in rather different ways. This observation not only applies internationally, but particularly within Papua New Guinea (PNG). This is not to suggest that there are no similarities to be found. Broadly speaking, host communities tend to undergo dramatic social, economic, political and cultural change, which includes the transformation of the landscape, new ways of understanding land and resources, social stratification as a result of the unequal distribution of mine-derived wealth, new political hierarchies and struggles, the loss or transformation of local practices, knowledge and institutions, challenges to traditional social relations, gendered impacts and new forms of economic dependency.

When we drill down to the local level, we find that these processes unfold in entirely unique ways. Local contexts, cultural complexities and histories decisively shape the nature of mining operations and the ways in which communities respond to change, but these factors can also determine the types of impacts and changes that might be experienced. Nevertheless, it still appears that the ways in which Lihirians have responded to the social and economic changes and impacts brought by mining are quite unusual when compared to other mining communities throughout PNG. I propose that a more specific — or a more cultural — explanation of the ways that Lihirians have dealt with the corporate mining economy can be found in their peculiar history of social movements. These activities, which peaked with the Nimamar Association in the early 1980s, provide a key insight into the diverse and diffuse ways that Lihirians have responded to mining. At the heart of these movements is a concern with social unity, morality and the attainment of a prosperous future. While these aspirations have endured throughout the colonial period and the more recent mining era, Lihirians have not always agreed on how they will be achieved.

A large part of this book is dedicated to understanding these movements and their recent manifestations, which can be regarded as discrete paths to particular destinations. Lihirians commonly gloss these as *ol rot* (roads) — a Tok Pisin metaphorical concept which refers to something between a cult and an ideology (Filer 2006). These roads sometimes appear to be mutually exclusive, but sometimes as interdependent. My particular interest is how members of a local political elite developed the Lihir Destiny vision, which charts a very specific 'road' to an idealised Lihirian modernity. Throughout the period from 2000 to 2007, these leaders were responsible for renegotiating the benefits package for Lihirian landowners and the wider community that was to be delivered by the mining company. This process provided them with the opportunity to develop the vision which is laid out in the Lihir Sustainable Development Plan (LSDP) — the latest benefit-sharing agreement between the company and the Lihirian community. This agreement, which we can call the 'Destiny Plan', is primarily a road map for a new future devised by local leaders who believe that their mandate is the pursuit of a Lihirian cultural revolution that will be facilitated by mine-derived wealth. However, their plans embody an inherent contradiction between the values of competition and cooperation: they want to create a wealthy egalitarian Lihirian society, sustained by entrepreneurial activities, that is free from the adverse effects of economic competition and the more general social malaise created by the mine.

The Destiny Plan is largely structured around an ideology of smallholder economics, which Lihirians commonly gloss in Tok Pisin as *bisnis* (business), or what Marx once called the petty commodity mode of production, or what the World Bank calls small-scale enterprise. This ideology was inspired by the Personal Viability movement, which seeks to develop an 'entrepreneurial class' throughout Melanesia. The Destiny Plan therefore raises important questions about the meaning of 'development' in Melanesia, particularly as its highlights the enthusiasms, ambivalences and contradictions in the various ways that Lihirians approach social change and large-scale resource development.

The counterpart to this ideology is the reification of traditional custom — or *kastom* as it is glossed in Lihir — which is seen as a particular road for social unity, the distribution of wealth, male authority and Lihirian identity in much the same way that the Destiny Plan is an alternative road for development and a reformed society. When Lihirians talk about *kastom*, they are usually referring to the performance of mortuary rituals that involve large-scale ceremonial feasting and exchange. In recent years, Lihirian *kastom* has grown significantly, partly reflecting the capacity of the ceremonial economy to absorb new objects and forms of wealth, while resisting the absorption of values and practices associated with them in the global capitalist economy. But this efflorescence is also due to a mass appeal to *kastom* for social stability which, like the earlier

Nimamar movement, can be understood as a direct response to extreme change in a compressed historical period. This particular road lies somewhere between an ideology of *kastom* and a kind of 'custom cult' which is made explicit through the performance of mortuary rituals, emphasising the ceremonial aspects of the local gift economy.

This process also raises some important questions about the nature of economic and cultural continuity. Lihirian culture has been vitalised by rapid industrialisation in a context where it might ordinarily be expected that the global forces of capitalism will have their most destructive impact. This recalls the classic work by Richard Salisbury (1962), who documented the efflorescence of traditional ceremonial exchange systems in the central highlands of New Guinea following the introduction of the steel axe. At the same time, we find that the articulation between the capitalist economy and the ceremonial economy is partly dependent on the nature of the local pre-capitalist social system, which reflects the pioneering arguments by Scarlett Epstien (1968) in her analysis of 'primitive capitalism' amongst the Tolai people of New Britain. These twin processes and their associated transformations, which are highlighted through the performance of *kastom*, lead to a further question about how mortuary rituals retain their significance in such an altered context.

The background against which these processes are played out is the corporate mining economy, which in itself has given rise to a further ideological construct, which Colin Filer (1997a) called the 'ideology of landownership', which holds that the key to development is the compensation paid by developers to the customary owners of natural capital. It is this belief which lies at the heart of Lihirian expectations for development and their dependency upon the mining company to fulfil these desires. At one level, the Destiny Plan fundamentally seeks to deal with this relationship, but at the same time, we find that the creative ways in which Lihirians use mining benefits in the performance of *kastom* actually assists them to carry their culture forward in times of change. Yet these are not simple processes or strategies; they are often highly contested, and have produced entirely unanticipated results.

If these different roads sometimes appear to be leading to different places, it is also true that they sometimes converge and cross over one another. Consequently, Lihirians seem to move within a hybrid cultural and ideological space that reflects the articulation of a corporate mining economy, a neo-traditional ceremonial economy, and a household economy of the kind that Marshall Sahlins (1974) would call the 'domestic mode of production'. My purpose in this book is to explore the ways in which Lihirians negotiate their way along these roads, and to understand the intersections or the hybrid space or the creative synthesis that is formed in the process.

Local World Histories

The changes experienced over the past century have prompted a deep local concern with social morality. The more recent ideology of *kastom* is underpinned by such concerns, often expressed in phrases like *pasin bilong Lihir* (the ways of Lihir). Many Lihirians often argue that their culture — which revolves around matrilineal men's houses, mortuary feasts and ceremonial exchange as institutions of nurturance, generosity and social reproduction — is fundamentally geared towards the expression of social morality or virtuous sociality. But even if Lihirians once exemplified the original 'virtuous society', there is a sense in which they have only recently become interested in such ideals.

These concerns, which are made visible through local discourses on *kastom* and development, are arguably a particular manifestation of modernity. Following Trouillot (2002: 225), who draws upon Koselleck (1985), I take the sense of radical disjuncture between the past and the present, and the perception that a different (and uncertain) future is both attainable and indefinitely postponed, to be among the most crucial elements of the experience of modernity — essentially a regime of historicity. From this perspective, Lihirians have been decidedly modern for some time. The changes brought through early interaction with the colonial labour trade, mission endeavour, sporadic engagement with the cash economy, the gradual disenchantment of their world, and the increasing discontent that arose from realisation of their marginal position on a national and global scale, are emblematic of this modernity. Aspirations for unity and morality dramatically increased as Lihirians grappled with the changes brought through colonial rule, Independence and later through mining activities. Social transformations challenged received cultural values and institutions; the desire for a reformed society emerged as Lihirians imagined a break with the past and looked to a new future.

World systems theory has found new life in grander theories of globalisation which assert that the global envelops the local, creating similarity in the place of alterity. The significant issue in this book is the construction of particularity in the face of apparently homogenising and universalising forces. The discourse of globalisation is often uninformative because of an over-emphasis on global homogeneity. Too often it is assumed that the globalising capitalist economy obliterates local economies, only to remake them in its own reflection. However, as we can see from the students' stories, regardless of the ways in which the Western epoch unleashes capitalism, 'it is always as an intertwining with local markets, paths of circulation, modes of production, and conceptions of consumption' (LiPuma 2000: 12).

Globalisation is often used as a loose term to imply the processes whereby economic and political activities in marginal places are shaped or dominated by distant developed nations and their linked transnational corporations. Other understandings emphasise the articulation of new forms of social organisation in an increasingly borderless world where flows of capital and new technologies propel goods, information, people and ideologies around the globe at unimaginable speeds (Lockwood 2004: 1). Even though mining has provided infrastructure, technologies, goods and opportunities that have transformed Lihirian lives and created a sense of 'interconnectedness', most Lihirians still remain on the periphery of the global network. Quite simply, villagers do not have access to the same technology and forms of communication that are taken for granted in urban areas. Although more information is available through new media resources, many Lihirians still lack the educational and experiential background required to make sense of these novel influences. Furthermore, given that community negotiations with the mining company are often based on misunderstandings about the price of gold and the prevailing belief that the company has access to inexhaustible funds which it is deliberately withholding from Lihirians, it is difficult to consider Lihirians as equally informed participants in any aspect of a global economic system (Macintyre and Foale 2004: 154).

Neo-Marxist interpretations of cultural change generally presuppose a depressingly inevitable conclusion whereby the 'civilizing' minority creates a world in its own image. World systems analysis and Marxist theory might appear anachronistic, but many people still think that, as the West materially and intellectually invades the rest of the world, peripheral peoples are merely the 'victims and silent witnesses' of their own cultural subjugation (Wolf 1982: x). Too often, analysis is directed at the ways in which people are encapsulated by the world system so as to 'suffer its impacts and become its agents' (ibid.: 23), rather than the ways in which they appropriate elements of this system for their own purposes.

To avoid what Errington and Gewertz (2004: 10) describe as the delineation of an inexorable and inevitable history, and to genuinely understand the process of cultural change, we need an anthropology that locates people within their own history, and that recognises the multifarious ways in which all people are agents of this history. This is an approach that emphasises the local articulations of global modernity, and that moves beyond simple binary oppositions between the West and the Rest, illuminating what Knauft (2002: 25) calls 'the social and discursive space in which the relationship between modernity and tradition is configured'.

In a place like Lihir, where Western capitalism apparently stands in opposition to traditional culture, it would be easy to revert to a dualistic approach. Many Lihirians do so themselves. Like a monument to the immorality and exploitation of capitalism, the mine processing plant towers over local men's houses and the 'moral exchange economies' that sustain social reproduction (Bloch and Parry 1989). In a Manichean showdown between the local and the global, it is only a matter of time before mining delivers the fatal blow that ensures Lihirians can do nothing more than recreate themselves according to an imposed image of civilisation. However, although the mining project epitomises modern technology, is the distilled essence of capitalism, and brings the differences between the local and the global into stark relief, we cannot simply equate this moment of neo-colonial capitalist expansion with Lihirian history as a whole. As Sahlins argues, 'it remains to be known how the disciplines of the colonial state are culturally sabotaged' (2005c: 486).

The Inevitability of Continuity and Particularity?

An adequate account of Lihirian lives within the context of mining must be approached from a temporal perspective that concentrates less on enduring structural relations and more on the relations between structure and time (Schieffelin and Gewertz 1985: 2). Marshall Sahlins takes an approach to history and anthropology which is guided by the premise that 'in all change there is continuity' (Sahlins 2005a: 9), and allows for a nuanced understanding of the relationship between continuity and change that shows how 'the transformation of a culture is a mode of its reproduction' (Sahlins 1985: 138). More importantly, his work reveals how continuity is itself an historical product, not the indication of a lack of history, for it is the result of 'happenings', not stasis. This viewpoint allows us to recognise the ways in which Lihirians have maintained their cultural integrity throughout a tumultuous period of change. Cultural categories shape moments in history and often endure in spite of change, even in those cases where the other player represents the dominant forces of the global capitalist system — whether it be the colonial administration or a multinational mining company. What emerges is an image of particularity that results from the interaction of internal social and cultural structures with the external influences that effect change. That is how we begin to understand Lihirian syncretism, for the use of introduced wealth and institutions has long been the systematic condition of their own culturalism — their authenticity and autonomy.

These processes are perhaps most economically captured in the concept of 'develop*man*' (Sahlins 1992), by which Sahlins integrates his opposition to the determinate hegemony of world systems theory with his ideas about indigenous agency and cultural integrity. The term was coined out of his mishearing of a Papua New Guinean Tok Pisin speaker's pronunciation of the word 'development'. Develop*man* is intended to capture that particular moment when indigenous peoples use Western goods (and institutions like those of capitalism and Christianity) to enhance their own ideas about life. Accordingly, his neologism is meant to reflect the 'indigenous way of coping with capitalism, a passing moment that in some places has already lasted more than one hundred years' (ibid.: 13). He argues that the first commercial impulse of indigenous people is not to become just like Westerners, but more like themselves: to build their own culture on a bigger and better scale than ever before (ibid.). It is precisely development from the perspective of the people concerned.

Develop*man* is easily recognised in many Melanesian societies: with the introduction of capitalism, there have been more pigs and shells exchanged than ever before, while new goods have made their way into the sphere of ceremonial exchange. Lihirian mortuary rituals are not insulated from the cash economy; instead, these practices are invigorated by access to new goods and forms of wealth. The so-called 'spectre of inauthenticity' (Jolly 1992) reflects Lihirian desires for modern lifestyles and the unanticipated paradox of our time: that globalisation develops apace with localisation (Sahlins 1999: 410).

As a follow-up to the bulk of his work on cultural continuity-in-change, Sahlins asks how this process might be ruptured. What is required to break this cycle, so that people will embrace Western values, make the achievement of develop*ment* the definitive goal, and thus become truly modern subjects? He suggests that the answer lies in cultural humiliation: people will not stop interpreting the world through their received cultural categories, and bending things to fit their values and categories, until they come to see their culture as something worthless. But before people abandon their culture, they must 'first learn to hate what they already have, what they have always considered their well-being', and to 'despise what they are, to hold their own existence in contempt — and want then, to be someone else' (Sahlins 1992: 23–4).

This is a critical observation. Lihirians have never (or not yet) passed through a phase of genuine cultural iconoclasm. Lihirians have long imagined a new future, the Destiny Plan is the latest attempt to realise this dream, and in some ways it does seek to hasten the transition from develop*man* to development. However, in each instance, different leaders have still imagined a grand, modern, or reformed *Lihirian* society. As we shall see throughout the following chapters, Lihirians have not suffered from any sort of 'global inferiority complex'. At times, they have felt frustrated, denigrated and marginalised, but their desires and strategies to achieve new lives for old have never hinged upon a kind of Faustian all-or-nothing bargain. Their search for development is not imagined as an absolute transformation, but rather as a total realisation. If people must first make a break with the past in order to imagine a new future — a conceptual objectification — this does not necessarily imply that they will then abandon everything from their past. Even though the concepts of develop*man* and development are probably better understood as ideal types (in the Weberian sense), they still provide a useful way to think through Lihirian experiences of modernity. As we look at the ways in which Lihirians simultaneously — or sometimes by turns — pursue develop*man* and development along the various roads on offer, then we begin to understand that cultural change takes place in the hybrid (and sometimes very uneasy) space that is created as people forge a new existence for themselves.

The Chapters

The chapters in the book are organised in such a way as to provide the reader with a background to the mining project, and then to trace the transformations that occurred from the colonial period through to the mining era. In the next chapter, I describe the history of the mine's development since exploration began in 1982, and the way in which the mine has come to dominate the local landscape and economy. In Chapter 3, I consider the history of Lihirian engagement with the outside world, outlining social and political developments that provided the genesis of social movements which prophesised an inverted world order. In Chapter 4, I provide a representation of Lihirian *kastom* as it was before the start of mining exploration, situating Lihir within a wider New Ireland 'areal culture'. Chapter 5 is built around the specific forms of social stratification that have arisen since the start of the mine construction as a result of the unequal distribution of mining benefits. Chapter 6 shifts the focus to the emergence of a local political elite and the Lihir Destiny Plan — especially the ways in which members of the elite have reinvented historical ideals in order to deal with the actual and prospective impacts documented in the previous chapter, including strategies for curbing supposedly 'wasteful' develop*man* ways. In Chapter 7, I return to the question of how mortuary feasts retain their status as the embodiments of tradition and *kastom*, and the contradictions between ideology and practice which reveal a deep-seated moral ambivalence about reciprocity, equality and exchange that threaten collective stability and expose the fantasy of virtuous sociality. In the concluding chapter, I summarise the argument that Lihirian cultural continuity and change is best understood through an examination of the complex intercultural zone which is created through the interplay of the local and the global.

2. The Presence of the Mine

Papua New Guinea has a long and turbulent history of mining activities reaching back to the latter part of the 1800s, when hundreds of Australians and Europeans came in search of gold on Misima, Sudest and Woodlark islands and on the Waria, Gira and Mambare (Yodda) rivers (Demaitre 1936; Healy 1967; Newbury 1975; Nelson 1976; Gerritsen and Macintyre 1986). While the scale and the impacts of these activities might not be comparable to contemporary large-scale mining, in most cases Papua New Guineans have hardly benefited in the same ways as foreign miners from the resources being extracted from their land.

The development of the massive Panguna copper mine on Bougainville in 1972 foreshadowed the dawn of an independent PNG in 1975. The mining exploration boom of the mid-1980s fuelled high expectations for economic benefits that would reduce dependency upon foreign aid and help create a more diversified national economy. It was hoped that the Panguna mine and the Ok Tedi gold mine that opened in 1982 would be the stepping stones to a prosperous future. However, the path to national deliverance via large-scale mining has been far rockier than anyone anticipated in the early 1980s. The outbreak of civil unrest in Bougainville in 1988 and the forced closure of the Panguna mine in 1989 became a national crisis as well as a local tragedy (May and Spriggs 1990; Denoon 2000). Meanwhile, international debates about the environmental impacts of the Ok Tedi mine grew louder, and an Australian court hosted litigation against Broken Hill Propriety Limited (BHP) as the mine operator (Banks and Ballard 1997). In the shadow of this turmoil, the PNG Government was moving on the development of another three large-scale gold mining projects at Porgera, Misima and Lihir which it hoped would support the post-colonial economy. Despite an abundance of resources, it has still proven exceedingly difficult to convert this natural wealth into wider economic growth, service provision and political stability.

Papua New Guinea has now developed a terminal case of the 'resource dependency syndrome' (Filer 1997b), manifest through two mutually contradictory conceptions of the role of 'resource rents' in national economic development: these revenues are either regarded as a form of 'economic surplus' which props up the state apparatus, or as a source of compensation — typically paid in cash — that fills the pockets of the local landowning population within the vicinity of each mining project. The strength and the appeal of this second conception owes a good deal to the development of an 'ideology of landownership' (Filer 1997a), which in itself was fostered by the processes of large-scale resource development. The exploration boom led many people to believe that their future share of development could be found in the various forms of compensation, services and infrastructure delivered by companies to the people who claim customary ownership over the resources being extracted. Modern citizens are thus 'customary landowners' entitled to receive royalties, rents and other benefits from the exploitation of their land. This image of entitlement is then confronted by national laws which posit state ownership of all subsurface resources.

Local and national reliance upon resource extraction is not unrewarding. In addition to the payment of taxes, royalties and compensation, mining companies also build local infrastructure and provide employment, services and business opportunities — the sort of development (or modernity) which the PNG Government has been unable to provide for most rural communities. Indeed, as we find in Lihir, mining projects conjure up fantasies and reconstruct myth-dreams in which the village is transformed into a city, and mining becomes the modern 'road of cargo'. While there is a popular tendency to paint multinational mining companies as evil monoliths without regard for local communities or the natural environment, the reality is far more complex. Every project within PNG has been profoundly shaped by the dialects of articulation between stakeholders with different agendas and levels of capacity to achieve their goals. The balance of power between companies and local communities might be uneven, but the relationship has never been unidirectional. As a result, the local impacts and responses, which are now well documented, are both varied and nuanced (Connell and Howitt 1991; Howitt et al. 1996; Ballard and Banks 2003; Filer and Macintyre 2006; Kirsch 2006).

At the Porgera gold mine in Enga Province, for example, relations between the company and the community are not only structured around competing expectations, but around the rapidly changing community dynamics that now threaten to destabilise both the project and the social viability of the Porgera Valley. The original inhabitants of the valley numbered around 2000 before large-scale mining began. It is now estimated that at least 40 000 immigrants reside within this area, and at least 1000 people from this group are regularly engaged in illegal small-scale mining activities within the official mining lease area (Callister 2008). The intractable problems generated by mass migration largely stem from the complex history of Porgeran social relations, networks and alliances. These are reflected in the cultural concept of *epo atene* — an Ipili verb construction meaning 'come and stand' (Golub 2005: 347), otherwise understood as a guest/host relationship — which makes it extremely difficult to exclude 'outsiders' or establish individual rights of ownership (Biersack 1999: 267).[1] This concept implies that people can reside or garden in any area provided they have

---

1   Essentially, guests may reside on land at the sufferance of the host who grants rights of usufruct and sojourn. Similarly, the institution of marriage within Porgera creates 'roads' and 'bridges' between descent

permission from the landowner, creating further implications for entitlements to compensation and expectations of things like relocation housing. The system is complicit in the influx of migrants who are eager to share in mine-related wealth and services, creating confusion about who are the rightful recipients of mining benefits, and anger amongst those who feel that they deserve exclusive — or at least priority — access to these benefits. [2]

Around the Misima gold mine in Milne Bay Province, which closed in 2005, local people now claim that they have very little to show for the sacrifices which they made. The strength of local customary exchange and feasting practices, and the enduring social ties between clan groups, meant that a significant amount of wealth was channelled back into these activities. While this brought certain socio-economic transformations and tensions, it also supported a broader level of social harmony (Callister 2000). At the same time, while Misimans have a long history of engaging with mining activities, and were aware that mine closure would eventuate, local landowners were unable to form cohesive allegiances to represent Misiman interests. Combined with a peculiar lack of company commitment to the social and economic consequences of mine closure, Misimans have been left with a divisive legacy as local leaders engage in expensive legal battles over the remaining trust accounts that were intended to provide for the future benefit of Misiman society. [3]

The lessons learnt from Bougainville and Ok Tedi, and the more general global shifts in extractive industry standards of best practice[4], mean that mining companies are now expected to be 'responsible neighbours' and contribute to equitable social and economic development. In PNG, operations are totally contingent on the capacity to maintain a 'social licence to operate', and so companies and government agencies have realised that the cost of doing 'business as usual' is now quite substantial. The Development Forum process, which was devised by the national government during negotiations for the Porgera mine in 1989, and subsequently applied to all other major mining and petroleum projects, has gone some way to ensure greater local participation in the planning process and the distribution of benefits (Filer 2008). This might create more involvement but has certainly done little to address the issue of dependency.

---

lines — idioms that reflect physical movement through marriage and the network of relational connections. The cognatic rule, which guarantees that all lines will intersect, is in stark contrast to unilineal systems; consequently there are no discrete, mutually exclusive groups in Ipili society (Biersack 1999: 263).

2   For further studies on the Porgera gold mine see: Filer 1999; Imbun 2000; Jacka 2001; Jackson and Banks 2002.

3   For further studies on the Misima gold mine see: Jackson 2000; MML 2000.

4   These standards are vast and varied, and include for example, international multi-sector principles like the United Nations Global Compact (UN 2000), and industry-led instruments like the Sustainable Development Framework of the International Council for Mining and Metals (ICMM 2003). International agencies are also influential in setting global benchmarks, such as the International Finance Corporation's Environmental and Social Standards (IFC 2006).

Furthermore, as we shall now see from the history of the Lihir mine, while the Development Forum enabled Lihirian leaders to develop a much broader definition of compensation in order to address immediate and future needs, it has not necessarily made it any easier to reconcile the competing interests and agendas between the different stakeholders, let alone to implement or achieve the multitude of their plans and visions.

Developing the Lode: An Overview

The Lihir gold mine is situated in the Kapit-Ladolam area of Aniolam Island, the largest island within the Lihir group. Unlike many other mining districts in PNG, Lihir does not have a history of alluvial gold mining. Gold traces were initially discovered during a geological survey of PNG conducted by the Bureau of Mineral Resources between 1969 and 1974. The results fuelled great expectations for substantial gold reserves on Aniolam. The report identified hydrothermal alteration and thermal activity on Aniolam, suggesting the possibility of an environment favourable to epithermal gold mineralisation. In 1982, prompted by these promising projections, Kennecott Explorations Australia and its joint venture partner Niugini Mining Limited employed geologists Peter Macnab and Ken Rehder to conduct sampling work on the islands which identified the potential for more extensive exploration. Rock chips taken off the sacred Ailaya rock in Luise Harbour yielded samples which averaged 1.7 grams of gold per tonne. Based on these results, Kennecott lodged an application for an Exploration Licence which was granted in 1982. Drilling commenced in the coastal area in late 1983, and continued into the adjacent Lienetz area through 1984. By the end of that year, the presence of a large gold resource had been confirmed.

Between 1985 and 1987, areas of anomalous soils in upper Ladolam Creek were sampled for gold, revealing the huge potential of the deposit. Drilling intersected gold values averaging 6 grams per tonne at intervals down to 197 metres below the surface. When the hole was deepened, a further 42 metres of gold mineralisation, averaging 3.92 grams per tonne, was intersected between 230 metres and 272 metres. This prospect was named the Minifie area and became the focus of diamond drilling throughout 1987. Further exploration defined several other adjacent and partly overlapping ore deposits, referred to as the Camp and Kapit areas. Kennecott engineers completed the first full feasibility study in 1988, but this failed to prove the economic viability of the project. In 1988, Rio Tinto Zinc (now Rio Tinto) acquired Kennecott from BP Minerals America and took over as the joint venture partner with Niugini Mining Limited.

A 'two mine scenario' was originally envisioned, whereby development of the Lihir project would be accompanied by the development of another mining operation on Simberi Island in the neighbouring island group of Tabar. While this would have posed significant logistical challenges, the more concerning scenario would have been the discovery of another orebody within Lihir, outside of the original licence area. While the Simberi operation eventually commenced in 2004 under the auspices of Allied Gold Limited, Lihirians have shown strong resistance towards further mining developments on their own islands.

The Lihir Joint Venture conducted almost nine years of test drilling, as well as detailed metallurgical test work, along with geotechnical, geothermal, groundwater, environmental and engineering studies. During this time, the Joint Venture devised strategies for mining in a geothermal area and removal of waste through deep sea tailings disposal. Following the submission of a final feasibility study to the PNG Government in 1992, and extensive community consultation, including detailed social and economic impact and baseline studies, between 1986 and 1994, the Joint Venture was issued with a Special Mining Lease (SML) on 17 March 1995. The lease is valid for the term of the company's Mining Development Contract — a period of 40 years.

In June 1995, Lihir Gold Limited (LGL) was incorporated in PNG for the purpose of acquiring formal ownership of the project from the Lihir Joint Venture. Four months later, on 9 October 1995, the initial public offering of shares was made. Lihirians secured 20 per cent of the overall 2 per cent royalty rate, and a 15 per cent equity stake through Mineral Resources Lihir Pty Ltd. Construction began in 1995, and by 1997 the processing plant at the Putput site was complete and the mine celebrated its first gold pour on 25 May 1997 at exactly 1.20 pm. Between 1997 and 2008, the annual gold production increased from 232 697 ounces to over 700 000 ounces. The project was initially operated by the Lihir Management Company (LMC), a wholly owned subsidiary of Rio Tinto, but the management agreement was terminated in October 2005 when Rio Tinto sold their 14.46 per cent share in LGL for A$399 million, in a decision to relinquish minority positions in other listed companies. Lihir Gold Limited itself became the operator, and by 2009, LGL was operating mines in PNG, Australia and the Ivory Coast in West Africa, and was producing a total of more than one million ounces of gold each year.

The Place for a Gold Mine

The Lihir island group lies to the east of mainland New Ireland, between the islands groups of Tabar to the northwest and Tanga to the southeast (see Map 2-1). The Lihir group consists of six islands (see Map 2-2): the main volcanic island of Aniolam, or 'big place' and, in order of distance, the low coralline islets of Sinambiet and Mando, and the three larger raised coral platform islands of Malie, Masahet and Mahur, collectively known as Ihot, or 'stone place', denoting the relative dearth of resources on these islands compared to Aniolam.

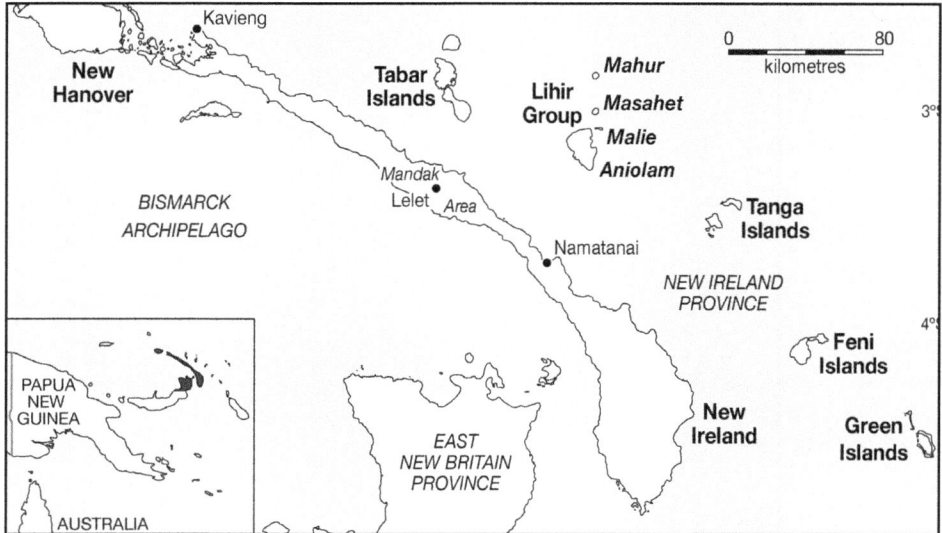

Map 2-1: New Ireland Province and Papua New Guinea.

When the prospect of mining in Lihir emerged in the early 1980s, there was a mixture of fear, ambivalence and enthusiasm. The historical experience of isolation and limited engagement with outside institutions meant that the arrival of the company was considered a major event. Although the initial fervour of expectation soon spread across all of Lihir, somewhat assisted by the belief among the members of the Nimamar Association that the arrival of the company was the fulfilment of past prophesies, it was initially the village communities of Putput, Kapit and Londolovit that were regularly engaged with exploration teams and involved in early negotiations. This was partly due to the location of the exploration camp at Ladolam on the shores of Luise Harbour (Plate 2-1), and partly to constraints on travel to and from other parts of the island group.

In the years preceding exploration, Lihirians were mainly settled in scattered coastal hamlets formed around local matrilineal descent groups. Most people relied on a form of shifting cultivation, combined with the partial domestication of a substantial pig population, and minimal amounts of fishing and hunting. Limited engagement with the cash economy through copra and cocoa plantations, small-scale business ventures, or full-time and part-time employment with the government or the mission, meant that the average per capita income across Lihir prior to 1983 was probably around K65 a year (Filer and Jackson 1986: 57).[5]

---

5   By the latter part of the 1980s the average per capita income was estimated at K100 per year (Filer and Jackson 1989: 90). The unequal distribution of income across Lihir was already noticeable in the early 1980s. Towards the end of the 1980s this inequality was further entrenched as more Lihirians came to rely upon the mining company for access to cash incomes.

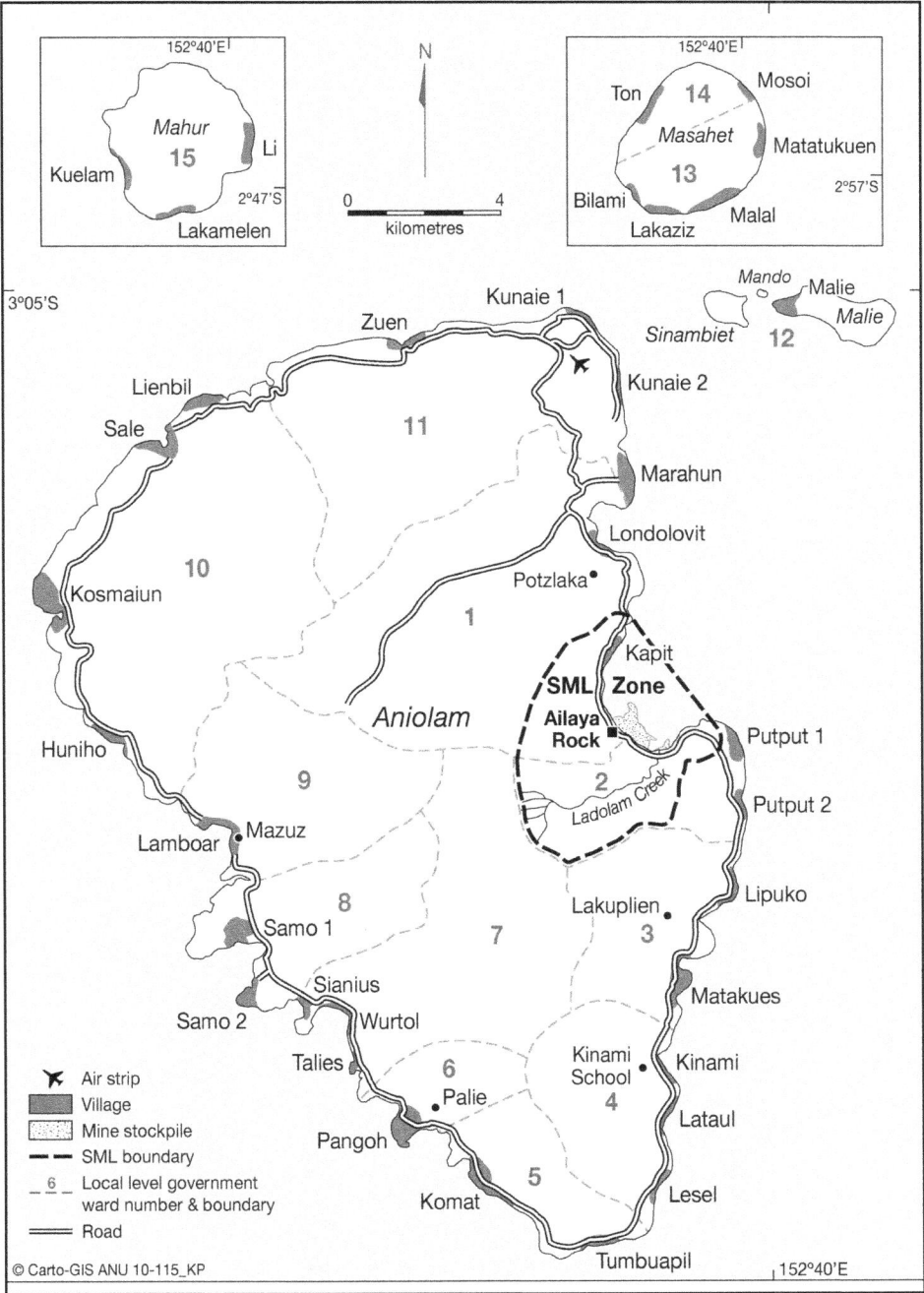

Map 2-2: The Lihir group of islands.

The Catholic mission station at Palie, on the southwest side of Aniolam, was the main service centre for the islands. It boasted a hospital, a primary school and a vocational school, while the mission boat, the MV Robert, provided access to the outside world. A vehicle track cut during the colonial period ran from Putput down to the southern tip of Aniolam and back up the western side of the island to Wurtol village, passing the wharf at the Palie mission station on the way. This ensured that villagers along the southeastern coast of the island had some involvement in early exporation activities. The northern villages of Aniolam were more isolated, both from each other and the outside world. Although Kunaie village was linked to the Potzlaka Patrol Post by a track that passed through the Londolovit Plantation, which was later designated as the site for the mining town, villages on the northwestern coast were completely cut off. Only a small number of dinghies and motorised canoes serviced Malie, Masahet and Mahur, which meant that their inhabitants were similarly isolated. By 1985, the people of Aniolam voiced a clear desire for the company to construct a ring road around the island that would connect all of their villages together. At that point, the only flights that made their way to and from the islands were those of the helicopter chartered by the mining company for its own purposes. The airstrip near Kunaie village was no longer in use, and it would be some years before the company thought it was necessary to build a new one to service its own operations.

Plate 2-1: Putput and Ladolam, circa 1985.

Photograph courtesy of the LGL archives.

Before the start of mining exploration, the population of the Lihir island group was thought to have increased from 3625 in 1925 to 5505 in 1980 (Filer and Jackson 1986: 32). There was a slight decline during the inter-war period, though not at the rate experienced in some of parts of New Ireland, such as the neighbouring island of Tabar. In the early 1980s, there were approximately 500 Lihirians absent from the islands. In the area that was to become the mining lease zone, there were an estimated 800 residents. When divided into three broad zones — North and South Aniolam and Ihot — the bulk of the population was found in South Aniolam, in a broad arc from Putput to Sianus. Kunaie and Londolovit were significant settlements in North Aniolam, with around 321 and 242 residents respectively, while the smaller islands supported a population of 1670. With the commencement of mining activities, many absentees returned to Lihir to seek new opportunities. By 1995, the Lihirian population had increased to 9892, and in 2007 it had reached 13 844 people. The non-Lihirian population has similarly expanded from fewer than 200 people in the late 1980s to an estimated 4000 people in 2008, in addition to approximately 2000 non-Lihirian employees residing in the camp and company housing. This sharp population increase is largely due to the return of expatriate Lihirians and to declining mortality rates as a result of improved access to health services. As we shall see in Chapter 5, over the life of the mine there have been significant demographic and settlement shifts associated with changes to the economy and service provision, and a growing non-Lihirian population.

Local Formations and Resolutions

By the time that exploration got underway in Lihir, mining companies operating in PNG were expected to commission detailed studies of the customary land rights relevant to their prospecting area. Negotiations for preliminary prospecting work had commenced on the basis of limited knowledge of Lihirian social structure and customary forms of land ownership. As early decisions and compensation payments were made, disputes soon emerged between the different landowning clans. Company and government representatives found themselves talking with individual clan leaders who either offered conflicting stories, or at least puzzling variations on the same story. The information offered in the original sociological baseline study prepared by a former District Officer (Smalley 1985) proved quite inadequate. By 1985, Lihirians had established a land demarcation committee. However this group was not equipped to deal with the rising number of disputes emerging from the new monetary value of the land, the inherently flexible nature of Lihirian land rights, and the tensions between different generations.

In 1985 Colin Filer and Richard Jackson were asked by the Lihir Liaison Committee[6] to conduct a social and economic impact study for the proposed mine, partly to address the shortcomings of the original baseline study. This report (Filer and Jackson 1986) was later fully revised and expanded (Filer and Jackson 1989). They found that Lihirians were already concerned about the future distribution of royalty and compensation payments and the potential for greater social division. Lihirians were familiar with the notion of compensation, and no one seemed to quarrel with the idea that it is due to the individuals and lineages who claim rightful ownership of the resources for which it is paid. However, royalties were something of a mystery. Some people thought that they might be equivalent to compensation for the use and abuse of agricultural land, while others regarded them as payment for the desecration of the land, the equivalent of a natural resource which should be free for all, as compensation for previous exploitation by the government, or even as the form in which volumes of cargo would eventually be delivered to members of the Nimamar Association (Filer and Jackson 1986: 99).

This confusion was exacerbated by the problem of distribution. The three prevailing ideas were that royalties could either go: (1) to members of the lineages that own the resources in the prospect area; (2) to all members of the clans to which these landowning lineages belong; or (3) to the entire Lihirian population, with each individual receiving an equal amount (Filer and Jackson 1986: 101). The difference between the first two options was obscured by the notion that 'clans own land' — an idea which the company had also come to believe, but which only partly captures the reality of traditional land tenure. At that stage, it was already evident that compensation payments were not being distributed throughout the entire clan. While it seemed logical that core landowners would prefer one of the first two options, there was widespread support for the third option, which was thought to be the least divisive, but was also based on some rather wild ideas about the amounts to be received. The reason for noting these early views is largely to demonstrate the changes that have since occurred. While Lihirian leaders eventually accepted the second option, believing that the nominated clan leaders would distribute their new found wealth equally throughout their respective clans, this has proven not to be the case, and the unequal distribution of royalties remains the biggest point of contention across the islands. Moreover, most Lihirians who do not belong to 'landowning clans' would still argue for a more equal distribution, but it is unlikely that any landowner would now contemplate an equal share for all.

In the early period of exploration, there was no unifying organisation representing the interests of Lihirian landowners. Lihir was largely divided between

---

6   The Lihir Liaison Committee comprised representatives from the New Ireland Provincial Government, the Department of Minerals and Energy, and the Kennecott-Nuigini Mining Joint Venture.

government supporters and the followers of the Nimamar Association, which was staunchly anti-government. At one level, Lihirians were united through their Christian faith, but again divided between Catholics and Protestants. Filer and Jackson recommended that Lihirians should form a 'council of elders' that contained representatives from each of the major clans in Lihir and the villages most affected by exploration activities, as well as representatives from the Nimamar Association, relevant government agencies, and the mining company. This group was not intended to replace the land demarcation committee; rather they would primarily deal with the land disputes arising in the villages of Putput, Kapit, Londolovit, and possibly Kunaie. In these early years, Lihirians were soon realising a distinction between 'landowners' and 'non-landowners': people with a connection to land within the prospecting area that would later become the Special Mining Lease zone, and those who are part of the unrecognised and undifferentiated hoard of collective custodians.

This council of elders never eventuated, but in 1989 Lihirians formed the Lihir Mining Area Landowners Association (LMALA). A young man called Mark Soipang, who was spokesman for the Tinetalgo clan which claimed significant portions of the prospecting area, was elected as its Chairman. Soipang and the LMALA assumed a leading role in future negotiations over the mining project, and in the following chapters, we shall see how both have maintained considerable influence across the political landscape.

As the Lihir Joint Venture prepared its final feasibility study, negotiations got under way for the relocation of Putput and Kapit villages. In March 1991, LMALA executives insisted that they should receive sitting fees and presented a list of eight demands to the company. When the company refused to agree to the demands, the LMALA imposed a 'stop work'. The history of negotiations has followed a similar pattern combined with a sense of mutual suspicion. Soipang has frequently stated his mistrust of Whites and has employed a diffuse range of negotiation tactics, which may well be the result of some rather strident advice he received back in 1988 from Francis Ona and his band of rebellious Bougainvillean landowners during a visit to the Panguna mine (Filer 1988).

In an attempt to address the mounting land disputes, a genealogical database for the proposed mining lease area was established, and local leaders discussed the need to codify the rules of traditional land tenure. In 1992, the mining company engaged Luke Kabariu, a Lihirian, to work with a combined committee including landowners, village leaders, village magistrates and government and company representatives, to produce the 'Lihir Land Rules' (*Lo Blong Graon Long Lihir*). This document never really reduced future disputes over the inheritance and ownership of land in Lihir, but rather highlighted the limitations of a simplified list of 'rules' divorced from the various ways in which these are enacted over the course of history. This work was carried out in tandem with the development of a Village Population System (VPS) database by John Burton, assisted by Kennecott liaison staff Martin Zanayes and Nick Ayen, which was in turn based on a survey of Lihirian men's houses by Colin Filer (Filer 1992a). The VPS database has been regularly updated over the life of the mine, and has proven to be an invaluable asset for monitoring social and demographic changes.

The Lihir Development Forum

On 1 November 1993, Lihirian leaders and provincial and national government representatives met at the Port Moresby Travelodge for a series of tripartite discussions known as the Lihir Development Forum. This forum provided an opportunity to secure joint endorsement for the project, and to produce a set of agreements between these stakeholders that outlined the costs, benefits, rights and obligations arising from the project (Filer 1995). Negotiations mainly revolved around the distribution of equity in the project, as Lihirians were pushing for an unprecedented 20 per cent stake. At this stage, there were already serious concerns about the political conflicts unfolding on Bougainville. New Ireland provincial leaders were determined to avoid a similar situation, and insisted that Lihirians should maintain a meaningful role in the management of the operation. Soipang presented the Lihirian perspective on this matter in a position paper which captured the ambiguous and strained relationship between Lihirians, the State and the project:

> The developers are foreigners and the State is only a concept. It is us, the landowners, who represent real life and people (Soipang in Filer 1995: 68).

While the State may be an ideological fiction, and the developers are certainly outsiders, Lihirians have still spent a great deal of energy working out how to extract benefits from them whilst reducing the extent of their influence. The protracted length of the Development Forum seems to have also arisen from the degree to which Lihirians felt alienated from the State — demonstrating the extent to which they have come to believe in it. By the end of 1994, this difficulty was exacerbated by the fact that Lihir was located within the electorate of the new Prime Minister, Sir Julius Chan, who would ultimately sign off on the mining agreement, but who had lost face with some Lihirians through his membership in the previous national government elected in 1992.

In Lihir itself, negotiations continued throughout 1994 on the issues of relocation, development of the plant site and town area, and the contents of the community benefits package. A road was surveyed and cut from Kunaie to Sale, eventually reaching Kosmaiun. Lihirian landowners visited the Misima gold mine in February 1994 to speak with local landowners and gain a better understanding of the transformations that Lihir would be likely to undergo. On 16 August, it was agreed that a Londolovit Plantation Development Planning Committee would be formed so that the traditional owners would have future input into the planning of the town and the use of their land. On the following day, the mining company handed the LMALA a draft Integrated Benefits Package (IBP) for review.

On 27 February 1995, the Development Forum was officially reconvened at the Malanggan Lodge in Kavieng, and most of the outstanding issues were resolved within two days. The provincial government representatives agreed that 20 per cent of the royalties would be paid directly to the landowners in cash, while another 30 per cent would be transferred to the Nimamar Development Authority (formerly the Nimamar Community Government), along with at least 30 per cent of the Special Support Grant to be used on community projects throughout Lihir.[7] The only issues not resolved in this meeting were the longstanding debates about equity and the distribution of a national government grant of K500 000 in start-up capital for new Lihirian companies. On 8 March, the National Executive Council (Cabinet) authorised the Governor-General to execute the Mining Development Contract between the national government and the mining company, and authorised the Prime Minister to execute the forum agreements with the provincial government and local community representatives. Three days later, Lihirian leaders refused to sign a separate IBP agreement with the company because they were still demanding a 20 per cent equity stake in the project.

The Integrated Benefits Package Agreement

On Thursday 16 March 1995, a delegation of nine Lihirian representatives flew to Port Moresby to meet with the Prime Minister and Mining and Petroleum Minister John Giheno and other government officers, in an attempt to finalise outstanding issues surrounding the IBP agreement, including the issue of project equity. The group included Mark Soipang and Ferdinand Samare, who was the Chairman of the Nimamar Development Authority, the national government's Lihir Liaison Officer Gabriel Tukas, and Ray Weber from the Lihir Management Company. Late into the night, Lihirian leaders finally accepted the government's offer of a 15 per cent equity share on the basis of the Prime Minister's promise that the overall royalty rate would be increased from 1.25 per cent to 2 per cent of the total value of output, and that the company would be able to claim the ring road as a tax deductible expense, which would save the Nimamar

---

7   The Special Support Grant is a transfer of money from the consolidated revenue of the national government to the provincial government hosting a major resource project. The grant has normally been equivalent to one per cent of the annual value of production from the project (see Filer 2008: 124, 1997b: 249).

Development Authority a considerable amount of money. These promises were confirmed in a letter from the Prime Minister and were attached as an appendix to the Memorandum of Agreement between the national government and the LMALA.

At 10 am on the following morning, the group attended the signing of the Special Mining Lease at Government House. This momentous occasion was attended by New Ireland Premier Samson Gila, and Provincial Minister for Mines Pedi Anis, and officials from Nuigini Mining Ltd. Speaking in front of the media and overseas bankers, Soipang said that Lihirians were 100 per cent behind the signing of the SML, but the signing of the IBP agreement must be carried out in front of Lihirian clan leaders back in Lihir.

Following the signing of the SML, arrangements were made in Lihir for the commencement of the Putput-Ladolam relocation. House blocks were pegged, and trees and resources counted. On Tuesday 4 April, a large feast was held at Putput to commemorate the move from the old hamlets to the new location along the coast. Soipang captured the gravity of the moment in his speeches during the event:

> Today we meet at this big occasion to remember the great change that will be part of the life of the relocatees. This great sacrifice of vacating the ancestral traditional land and going through the process of relocation to new land was a major decision for the relocatees. The decision to relocate was not an easy hurdle. The decision they made was not merely for their good alone. Rightfully, it was for the benefit of the Lihir project, Lihirians, New Ireland Province, PNG and the world. ...

> We on Lihir, New Ireland and PNG have much link to the land and therefore to leave your land and especially when you are forced to leave and relocate to a new land is a great inconvenience. ...

> Relocation has opened the door for the mining company to operate. ...

> This village will no longer be the same as it is today. The hausboi at Maron, Kabanga, Latawis Lamabarai and the Putput cemetery will be preserved. The physical nature of this land will change to allow the construction of a wharf, plant and houses. After the present state of the land has been changed, we will not return it to its natural state. ... We will not forget this day because we will continue to remember that we were the last group to gather at this feast before the natural state of the land was changed (quoted in the *Lihir Gold Times* 5(2), 1995).

The final ceremonial signing of the IBP agreement was set to take place in Lihir on 20 April. At the last minute, the affair was delayed because Lihirian leaders had not been able to acquire the desired 400 pigs for the event, and also because they were still unsatisfied with the content of the agreement. On 26 April, the event finally proceeded. It poured down throughout the day and reportedly a young girl drowned in a rip-tide while adults were preoccupied with the proceedings. Filer (1995: 69) suggested that all of this might be regarded as a 'bad omen' or a sign of worsening things to come. Many Lihirians might now find now themselves in agreement, but at the same time I would still be hard pressed to find people who want to completely turn back the clock.

The broad sense in which Lihirian leaders approached the issue of compensation — particularly by comparison to those of other landowning communities — was largely explained by the Development Forum process and the belief that they could make the compensation package work for Lihirians. During the ceremonial signing of the agreement, Soipang outlined the ways in which members of the local political elite were approaching this issue:

> There have been times set to sign the heads of agreement but we have long deferred the signing until we finally decided to sign today. We have to delay the signing because we would like to freely make the decision to commit life, land and our environment to the project. ... the negotiation process took many years until a final agreement was reached. The landowners made a lot of compromises and for that reason we are ready to sign the agreement today. ...

> In the negotiations the landowners had four objectives to achieve:

> 1. We strive to maintain that during the operation and development of the Lihir Gold Project, development must be a true development in the life and welfare of the people in every village of Lihir. This means that the development of the mineral resources must go hand in hand with development of the human resources.

> 2. We would like to see a balanced development that will extend to all areas of Lihir in order to maximise the gap between privileged and non-privileged Lihirians.

> 3. We would like to see a sustainable development in which Lihirians will care for and maintain.

> 4. We would like to maintain that at all times there must be stability, peace and harmony within the Lihir group.

> We aim to see that these four objectives must bear fruit for the development of the Lihir Gold Project in the Lihir group.

> The strategy we will follow to achieve these objectives is:

1. The structure of the Nimamar Development Authority is based on the Lihir Clan System.

2. There is already in place the Lihir Society Reform Initiative and Programme.

3. The Village Development Scheme is included in the Compensation Agreement.

The principle of the Integrated Benefits Package in the compensation agreement clearly outlined that, in destruction, payment goes direct to the immediate landowners. Development is development for all areas and the people on Lihir. Security is an investment to cater for development and Rehabilitation is a process of reclaiming land to make it once again suitable for people to live on.

The landowners, having the knowledge that all their land within all mining leases will be destructed and hence will lose many benefits for the good of the Lihir Gold Project, Province and Country and the World. This is a great sacrifice the landowners will endure and therefore are prepared to take it because it is for the good and benefit of all the people.

Finally, the end of the hard negotiations is over and so we are now ready to sign all the agreements to allow for the development of the Lihir Gold Project in order to allow for developments in all areas and within the people of Lihir; to create development in the province, country and the world (quoted in the *Lihir Gold Times* 5(2), 1995).

Local leaders were clearly searching for the right formula to maintain social stability and generate positive social changes. The earlier visions of a new society were now distilled in a model of 'balanced' development to be funded through the IBP and supported by something called the Society Reform Program that was supposed to protect Lihirian custom. This was the genesis of the later Lihir Destiny Plan that imagined the revolution in Lihir as key to the broader transformation of the country.

The IBP agreement was divided into four chapters: Destruction, Development, Security, and Rehabilitation. It contained a set of separate agreements between the LMALA, the Nimamar Development Authority (which became the Nimamar Rural Local-Level Government in 1997), the Lihir Management Company and the Government of PNG. This package outlined the specific range of agreements and memoranda that cover different aspects of mine-related development, including compensation, housing relocation, infrastructure commitments by the local and national governments and the mining company, and commitments relating to environmental monitoring. Special Mining Lease block executives were entitled to a 20 per cent share of royalties in addition to compensation for damages and destruction to resources, and Lihirians gained a 15 per cent equity share in the project through shares purchased on their behalf by Mineral Resources Lihir Ltd. In addition to compensation distributed throughout the 'affected' area and the general construction of infrastructure throughout the islands, the IBP also contained provisions for a Village Development Scheme (VDS). Funding was made available through this scheme to provide Lihirians with assistance for housing improvement and basic community infrastructure such as reticulated water and electricity, and sanitation services — the trappings of a modern village existence. Importantly, there was also provision for the agreement to be reviewed every five years. Although the IBP was the result of nearly a decade of stakeholder negotiations, and became 'the new benchmark within Papua New Guinea for such arrangements' (Banks 1998: 62; see also Filer et al. 2000: 53–8, 71–6), it is still arguable that the only genuinely integrating factor was the blue binding holding these separate agreements together.

While the Development Forum provided local leaders with a seat at the negotiation table, and has assisted greater self-determination, we must recognise that the forum process — and the industry of which it is a part — is still structurally geared towards dependency. It is effectively an opportunity for landowners to present a wish list to the government and the company, who are then held responsible for delivering these dreams through a series of agreements. In return, landowners just have to promise to be well-behaved and not disturb operations. The pages of government and company commitments outlined in the IBP are set against the single Lihirian commitment to 'cooperate'. Granted that there is some moral justification for this imbalance in the fact that landowners make substantial sacrifices, and that the benefits which accrue to the company far outweigh anything received by the community, this arrangement still fails to address the difficult task of converting resource rents into long-term development. It may even render the Lihirian desire for sustainable or 'balanced' development completely futile, especially since there are no real repercussions when men behave badly. It is for these reasons that the new directions proposed in the revised IBP, the Lihir Destiny Plan which I consider in more detail in Chapter 6, appear all the more remarkable. The determination to achieve 'balanced' development from mining, through an emphasis on individual responsibility and *bisnis*, not only entails plans for the post-mining era (something which is often absent at other projects), but also marks a significant departure from previous approaches to mining and community development.

Relocation and Construction

Throughout 1996 and 1997, there were a number of significant developments that provided new services and connectivity, and also permanently altered the landscape and existing lifeways. On 21 January 1996, the old Ladolam exploration camp was decommissioned and workers were shifted to the Putput construction site and a new construction camp on the Londolovit plateau. Around 20 government houses and 70 company houses were constructed, as well as a large abode for the Chairman of the LMALA, new accommodation for the majority of the workforce who would operate on a fly-in-fly-out basis, a hotel, a sporting club, and an international primary school. A modern hospital was built to service both employees and the wider community, and in the local townsite work commenced on a new police station, market place, a shopping district with banking and postal facilities, as well as offices for the provincial government, the LMALA and the LMC Community Liaison Department. The area leased from customary owners — which excludes Londolovit township, portions of Luise Harbour, and other areas previously alienated such as church blocks or government land — amounts to approximately 2700 hectares, roughly 15 per cent of the surface area of Aniolam that remains under customary title (Filer and Mandie-Filer 1998: 3). The ring road was finally completed and opened by Prime Minister Sir Julius Chan on 25 May 1996. It stretched over 74 kilometres, connecting the mine, the township, the mission station, and some 20 coastal villages. A new airstrip was also constructed at Kunaie and opened with a feast on 17 April 1996. On the following day, 16 landowners (including one woman) flew to Port Moresby and back on a Dash 7, marking the commencement of commercial flights.

Under the terms of the Putput/Ladolam and Kapit Relocation Agreements in the IBP, some 215 people were listed as residents to be relocated to make way for the mine. Prototype relocation houses were initially built to give people an idea of the style of housing on offer. These designs were more environmentally suitable, but they lacked the familiarity of the high-set weatherboard houses commonly used for government employees. So the original designs were abandoned for a more recognisable and acceptable style. At the time, these houses represented current ideas about what it meant to be a modern, wealthy homeowner. These houses were seen as the preserve of middle class businessmen, bureaucrats and administrators, and signified access to a world of relative luxury and wealth previously denied to Lihirians.

In accordance with the relocation agreements, the LMC agreed to provide two types of relocation house for those residents of the affected areas who had established their identity as customary landowners. Type A houses were three-bedroom high-set 'deluxe' model homes (see Plate 2-2). These houses were given to family heads with land in the pit area. The company built these houses at no cost to the resident and provided K200 per annum for maintenance, plus a one-off K2000 furniture allowance. Type B were low-set two-bedroom homes with similar appearance but fewer features. Occupants were to receive a smaller furniture allowance of K1000 and a maintenance allowance of K200 per annum. The choice between the two largely depended upon the actual or anticipated size of the household to be relocated. Members of existing households (mostly adolescent children of older landowners) were also to be provided with a kit of materials to construct a small Type B house, either three years or five years after the agreements had been signed (Filer et al. 2000: 72). In addition, one grave was relocated and seven new men's houses were also constructed by the company to replace those lost in the relocation.

Plate 2-2: Newly constructed relocation house in Putput, 1995.

Photograph courtesy of the LGL archives.

The company managers recognised the need for a women's section in their Community Liaison Department, and before construction started, they appointed a consultant (Susanne Bonnell) who assisted Lihirian families in their domestic transition and was instrumental in the initial moves to set up a women's association. In 1995, when Martha Macintyre was engaged by the company to set up an annual social impact monitoring program, she worked closely with Lihirian women to establish the Petztorme Women's Association (Macintyre 2003). While this has provided women with a voice in a predominantly male political environment, the organisation has been consistently hampered by sectarian politics, which has muted their involvement in the development process.

From the outset, the relocation program divided Lihirians, and in many ways set up a new hierarchy based upon individual distance to permanent environmental destruction. Residents of Putput 1 village were relocated to nearby land around the corner, which meant that they could retain a semblance of village unity. This new village was known as 'Relocation', and later called the new Putput 1. The unrelocated part of the old Putput 1 village was reclassified as part of the existing Putput 2 village. Reticulated water and electricity were initially delivered to Putput 1 (or Relocation), and new modern men's houses were constructed from permanent materials. This was later extended to Putput 2. Most other Lihirians have subsequently conflated the two village areas and simply refer to 'Putput', which is generally considered a kind of 'landowner suburb'. While Putput residents are noticeably better off compared to most other Lihirians, and Putput is now a decidedly 'urban village', there is a high degree of internal stratification and conflict within this core landowning area.

Kapit residents were relocated shortly after Putput. However, because they did not have land nearby which was not intended for mining purposes, they were not relocated as a single residential community. Instead, they were scattered around Lihir to places where they had clan connections. Some shifted to the neighbouring village of Londolovit, while most moved away from the growing centre of economic development to villages on the western side of Aniolam, such as Talies, Wurtol and Kosmauin. In 2007, some Kapit people were still refusing to be relocated due to a lack of suitable land, reluctance to depart from sacred sites, and political struggles over compensation. Kapit and its famous hot springs have since become a caustic stench on the mine's political landscape.

The Ailaya

It should come as no surprise, perhaps, that the original rock chips sampled from the sacred Ailaya rock pinnacle, rising out of the central caldera in the Luise Harbour (Plates 2-3 and 2-4), have precipitated the profound transformation of the Lihirian lifeworld.[8] Properly understood, the Ailaya rock is a portal to the land of the dead for all Lihirians from across the group of islands. The spirits of deceased Lihirians enter this land of the dead, known as *Rondende*, through a central marine passage at the base of the Ailaya, assisted by the spirits of relatives already resident there. In Lihirian conception, the distribution of the islands reflects a particular cosmogonic event — an original eruption of the central caldera on the main island of Aniolam, throwing out the islands of Malie and

8   There have been various spellings of Ailaya reflecting different local dialects, such as Ilaia, Alaia, and Ailaia. In 2009, the Lihir Cultural Heritage Committee decided upon the spelling Ailaya. The 'y' is intended to symbolise the Y-shaped entrance to Lihirian men's houses, where clan members are often buried (see Chapter 3). In the words of Peter Toelinkanut, 'the Ailaya is the gateway for all Lihirians'.

Sinambiet. Formerly there was a smaller rock pinnacle adjacent to the Ailaya, alternatively called Tuen kanut or Ai tuan tamberan, which can be translated as 'ghosts bones'. Spirit beings, or *tandal*, which comprise the wider sacred geography of Lihir (see Chapter 4), are held to have emerged originally from the Ailaya site complex, before fanning out across the Lihir group and taking up residence at their respective sites across the islands. To the extent that there is an explicit Lihirian eschatology, it is couched in terms of an ultimate, centripetal return of the *tandal* to the Ailaya, in a reversal of the original centrifugal event.

Plate 2-3: The Ailaya, circa 1985.

Photograph courtesy of the LGL archives.

Plate 2-4: Drill rig and helicopter on the sacred Ailaya rock in Luise Harbour, circa 1983.

Photograph courtesy of the LGL archives.

During the initial mortuary rites that honour deceased male leaders, and in some cases senior women, the deceased are decorated with red clay and shell valuables and seated within the confines of their men's houses in preparation for burial. Clan members gather in the men's houses and mournful *rangen* songs are sung throughout the night. The ultimate purposes of this ritual are to farewell the deceased and assist them in their journey to the afterlife. On Aniolam it is understood that the soul of the deceased walks across the island or along the coast to the Ailaya, resting upon the Tuen kanut before reaching its final destination, *Rondende*, before sunrise. On the outer islands of Malie, Masahet and Mahur, the spirit of the deceased must first journey across the sea to reach Aniolam where the Ailaya is located. This journey is symbolised by seating the deceased in a special canoe called a *konkonla*, placed within their men's house (Bainton et al. forthcoming a).

On a regional Pacific scale, the Ailaya is an exceptional site. Many communities in the Pacific have held notions of lands of the dead, but for the most part these are abstract spaces, usually entered via the sea, or located in valleys or on distant peaks. Throughout New Ireland, similar places are located out to sea, on the horizon, or on another island (Bell 1937: 330; Clay 1986: 50; Kramer-Bannow 2008: 195). The physical presence at Lihir of a portal to the underworld is exceptional, and no doubt reflects the unique form of the Ailaya and its setting within an active volcanic landscape. More unusual still is the inward-focused cosmography of Lihir, in which the Ailaya serves both as the geographical pivot and the point of temporal initiation and closure.

The Ailaya now sits in the middle of the SML area (Plate 2-5) and acts as a cap to a rather large lode which the company has long sought to mine. While the Ailaya has thus far been preserved through an agreement between landowners and the company, it has suffered numerous disfiguring events, including the cutting of benches into its crest to facilitate early test-drilling and the development of the first coastal road along its sea front. Cut off from the sea by a widening corridor of rubble and overburden supporting a series of vehicle access roads and pipelines, the Ailaya is now a forlorn remnant of its former shape, a silent island in a churning sea of mining activity.

By virtue of its position within the SML, the cosmological significance of the Ailaya has since been overshadowed by its political significance. Although the shared relationship to the Ailaya once entailed obligations of reverence rather than rights to specific uses (Macintyre and Foale 2007: 53), the prospect of royalty payments from the mine, which was already thought likely to destroy the site, quickly generated complex arguments among Lihirians over the ownership of the Ailaya. Throughout the life of the project, independent risk analysis and social impact studies have consistently recommended that the Ailaya should not be destroyed or damaged, and local leaders have maintained the importance of the Ailaya to Lihirian social harmony.

On the 1 October 1984, Soipang lodged the first compensation claim on behalf of the Tinetalgo clan for damages to the Ailaya site (Kabariu n.d.). No figures were mentioned and there are no records that indicate whether any payment was made. On 28 August 1985, Soipang announced that the Tinetalgo clan would perform a customary feast as compensation for the destruction of the area where the *tandal* known as Kokotz resided. This was the first time that the Tinetalgo clan had staged a feast for the *tandal*, which some considered to be a tactic for the Tinetalgo clan to gain sole rights to the sacred area.

Plate 2-5: The Ailaya in the middle of the mining pit, 2008.

Photograph courtesy of the LGL archives.

The politics of ownership intensified when construction started on the ring road. The only feasible route through this section of the island was around the base of the Ailaya, meaning that part of the sea frontage of the rock would have to be removed. During a community meeting held on 22 March 1989, with people from Putput, Kapit, Londolovit and Wurtol, the majority of people agreed that the road could commence. Soon afterwards, a younger group representing the Tinetalgo clan, led by Soipang, disputed the decision and demanded compensation. The Mining Warden was called in, and he recommended that a compensatory feast should be held and that Kennecott should provide the pigs and food. This was rejected by the Tinetalgo clan, whose members argued that this would effectively grant Kennecott part ownership over the Ailaya.

As outlined in the IBP, the LMC paid K50 000 compensation for damages to the Ailaya which was received by the Tinetalgo clan. The clan 'distributed' this money by hosting a large feast at Sianus from 26 to 27 June 1996, which was supposed to mollify the Kokotz *tandal*. The LGL Board of Directors attended, but many Lihirians boycotted the event. Even if this feast did not confirm ownership, it did provide the Tinetalgo clan with the right to claim any future compensation payments for damages to the area. Few Lihirians accept that the Ailaya can be owned by any one group or individual. Consequently, most Lihirians have removed themselves from debates about the Ailaya to demonstrate their dissatisfaction, which has only further alienated the Tinetalgo clan from the wider Lihirian community.

In the IBP, the LMC also agreed to the establishment of a committee to review the options for restoration of the Ailaya. This committee would consist of representatives of the Tinetalgo clan, the wider Lihir population, and the LMC. The original group was called the Ailaya Restoration Committee, although the members later changed its name to the Ailaya Preservation Committee, firstly because they thought that the rock was now beyond restoration, and secondly to suit the struggles and disputes over the boundary of the Ailaya area. Internal politics, possibly related to the high representation of Tinetalgo clan members, has precluded any meaningful action and the committee is now defunct.

Despite its remarkable cultural significance and the commitments made in the IBP, company managers have consistently entertained the possibility of mining the Ailaya, euphemistically referred to as 'coastal mining'. Soipang has always vocally defended the Ailaya, though perhaps more recently for political reasons as a bargaining chip with the company. Company managers are now sceptical about the contemporary relevance of the Ailaya, further convincing themselves that the economic benefits to be gained by Lihirians (and LGL) from mining the area far outweigh any spiritual significance it might have. Managers have seized upon the realisation that some Lihirians would certainly agree with this argument, especially younger or disgruntled community members. However, this argument fails to recognise that for older Lihirians the total destruction of the Ailaya is perhaps an unfathomable event with cataclysmic repercussions. Moreover, future generations who do not feel so sanguine about the loss of this sacred site may well consider its destruction as outright sacrilege and grounds for legal action.

Islands in the Global Stream

Mining has brought unprecedented material, cultural and political change to Lihir. While many people have been disappointed or frustrated by what they perceive as insufficient or inadequate change, the reality of this transformation is fully known in comparison with other parts of PNG where people are still 'waiting for company' (Dwyer and Minnegal 1998). Managers like to imagine that the impacts extend to the limits of the SML zone, and that Lihirians should be grateful for any benefits that flow beyond this boundary. However, the size of the mine is not proportional to its presence. It is this focus on the physical nature of the mine that seems to keep managers from recognising that the mine actually sits within the middle of the Lihirian community and completely occupies their lives.

Lihirians now have access to a variety of media sources and telecommunications, and there is a daily flow of Lihirians and non-Lihirians on and off the island who act as important conduits for information about the outside world. The national papers are delivered daily, TV services exist in villages close to the mining area, and in 2007, Lihirians entered the digital age as mobile phone technology reached the Londolovit town site. In short, mining has already brought a lot of the material, cultural and temporal features of 'modernity'. Lihirians might not be equally informed participants in the global economic system, but compared to their neighbours and to pre-mining days, Lihirians have become decidedly more connected and exposed to the lifestyles and cultural visions of others very different from themselves.

While the incorporation of new forms of agricultural production occurred gradually through the colonial years, and entailed the extension of primary subsistence activities, mining has introduced industrialisation instantly and on a grand scale. In a matter of years, Lihirians have been confronted with a range of significant changes to their lives, including the dramatic transformation of the landscape; the introduction of industrial wage labour relations; the massive influx of foreign workers, as well as the marked shift towards modern technologies and techniques. Economic dependency on the mine is not only evident in the demand for compensation, but also in the extent to which local businesses rely upon the presence of the mining company, and the ways in which the ceremonial economy relies upon the incorporation of mining benefits and wages, which has sustained *kastom* through this revolutionary period. Quite simply, mining has pervaded all aspects of the Lihirian lifeworld, and their engagement with capitalism has occurred on a wholly different scale to most other Melanesians.

Although Lihirians have been criticised by government officials, other Papua New Guineans, and even some of their own leaders, for being 'wasteful' or failing to capitalise on their moment, this is not solely due to the structure of mining agreements or because Lihirians are 'maladapted to modernity', but rather because particular historical circumstances, and the scale and speed with which industrialisation has been introduced, have resulted in spatio-temporal disjuncture. Lihirians have been thrust into the modern world, but they have distinctly shaped their trajectory. In the following chapters, we shall look at the ways in which Lihirians have historically responded to change and conceptualised the source of that change, and at the received cultural orientations that have influenced their engagement with new political and economic systems and shaped the outcomes of this remarkable experience.

3. *Las Kantri*: Lihir Before the Mining Era

Then I saw a new heaven and a new earth, for the first heaven and the first earth had passed away.... (Revelations 21:1).

All of our talk was coming up. Those who died before, now they had come up, they had come up and we recognised them. But they couldn't talk to us and tell who they were, but we saw them and knew ... we recognised their walk, their eyes, their head ... now if you have a mark and you die, we can still recognise you.... When Kennecott was here, if you were hungry, go and eat first, you don't need to buy anything ... we took all sorts of White man's food, and we did not buy it. Food wasn't anything, if you were hungry, you go and eat, you just go. Everything was free ... Arau's talk was true, the ships were full up. Later he said when the ships come here, money will pour out on us. Every house will be full up of things belonging to White men. Now all the men started to look and they believed him now.... When Kennecott left so did all of our ancestors. Who will make these things come up now? Arau had said that everything will be free for us, but they [Kennecott] went back now and now things were not free, everything was hidden again (Bah Arom, Kinami village, Lihir, 2004).

This chapter offers a short economic and political history of Lihir to illuminate historical influences on the ways in which Lihirians have responded to mining, and to illustrate the genesis and escalation of desire for economic and political autonomy (not necessarily outright secession), and the rising antipathy towards the colonial administration and later the State. Earlier social movements — namely the Tutukuvul Isakul Association (TIA), which evolved into Tuk Kuvul Aisok (TKA),[1] and later the Nimamar movement — are the combined result of moral inequality between Lihirians and Whites, and the gradual process of pauperisation that encompassed Lihir under the Australian administration, generating a sense of economic frustration. These movements preached a form of millenarianism whereby Lihir would become the *las kantri* (last country) — the utopian new world. By tracing political and economic developments through the colonial period, particularly the latter years of the Australian administration (1945–75), the peak of these social movements (in the 1960s and 1970s), and later developments in the 1980s that led up to the beginning of mining, I provide a genealogy of Lihirian marginality, discontent and desire for wealth

---

1   *Tutukuvul isakul* means 'stand up together to plant' in northwestern New Ireland languages; *tuk kuvul aisok* means 'stand together and work' in the Tigak language of that region.

and independence. This reveals not only the conditions under which these movements emerged, but also their defining characteristics that have continued to influence Lihirian praxis and ideology. Thus, a historical understanding of the ways in which Lihirians shaped and understood the colonial project and national independence tells us something about the contemporary political and economic climate in the context of large-scale resource extraction.

This chapter is intended to reflect the ways in which Lihirians have collectively reconstructed their sense of historical time. The 13 years of mining exploration and negotiations have been condensed into a revolutionary moment of change — a single historical cleavage. It is not that people fail to recall different moments throughout this period. Putput residents especially remember the mixture of anticipation, excitement, trepidation and sadness that accompanied early interactions with exploration teams, drillers, government and company representatives, the establishment of the Ladolam camp, employment opportunities and new wealth, the embattled negotiations and the experience of relocation. And while most other Lihirians were not so closely aligned with the project, many still recall a steady stream of political, economic and social changes in their lives. However, the combination of lapsed time and profound change means that, in popular discourse and memory, history is now often divided between mining and pre-mining periods. Many Lihirians think that the present — *nau* (now) — is spiralling into a state of social anomie which they compare with an Edenic, ordered and stable past — *bipo* (before). Social history has been divided in ways that essentialise the past and the present as dichotomous states.

Lihirian historical accounts are deeply imbued with a 'structural nostalgia' (Herzfeld 1990). It is often in people's political interest to minimise the perception of historical change prior to mining and to develop narratives of unprecedented rupture imposed on an unchanging social environment. Although many Lihirians persistently present an image of past stability, cohesion, morality and order, Lihir has long been experiencing change, and there is scant evidence that Lihirians were not always subject to mutability and the inherent instability of local political formations. Without suggesting an earlier time of harmonious social cohesion, any 'stability' that might have once existed was purely the result of leadership qualities that were still 'relevant' to the social and economic climate.[2] Each generation required new leaders equipped to deal with new changes. However, large-scale resource extraction invariably produces change at a faster rate than that to which local communities (and their leaders) can adjust

2   Throughout much of PNG, disputes, conflict or even instability, are not symptomatic of social anomie, rather they are a relatively normal state of affairs. The pacifying efforts of the colonial administration may have successfully targeted endemic warfare throughout much of the Territory, but as Goldman (2007: 69) argues, 'even for the most sensitive of post-colonial consciences, a "community without conflict" is neither a destination objective, nor a socially imaginable outcome'.

(Jackson 1997: 106–7). The recent shifts in Lihir highlight how pervasive and detrimental the effects of capitalism can be, even on the most flexible of social structures, especially when it arrives through large-scale mining.

Notwithstanding the comparatively sudden impact of mining, the changes that took place in Lihir over a much longer duration through the twentieth century were equally pronounced. The ways in which Lihirians responded to external influences from missionaries, merchants and administrators reflected comparable colonial encounters occurring throughout Melanesia and other parts of the world. Although social scientists have long held a deep fascination with indigenous responses to externally imposed change throughout the world, Melanesian responses have generated some of the most sustained anthropological attention. Of particular interest was the proliferation of spontaneous local movements from the 1950s onwards, that differed in their specific objectives and origins, but shared broadly similar concerns with the achievement of economic, social and political development through forms of communal action. The bewildering variety of movements, which the colonial administration loosely termed 'cargo cults', differed according to local historical experiences, customs and epistemologies.

Many of these movements contained an uneasy combination of local mythology, eschatological and subaltern theologies, ritual, communistic sentiments, new forms of leadership and economic activities (often centred upon cooperative efforts), prophesy, visions and supposed miracles, material objectives, and a deep concern with moral equality (Lawrence 1964; Worsley 1968; Lindstrom 1990; Burridge 1995). Both the colonial administration and national politicians have used the term 'cargo cult', often pejoratively, to describe any seemingly bizarre behavior or local political activities which often have little to do with cargo expectations as narrowly defined (Walter 1981). However, clear differences can be found among these movements and responses. Some were overtly political, and are better described as micro-nationalist movements (May 1982); others resembled self-help movements or development associations (Gerritsen et al. 1982), and some were decidedly 'other-worldly' or millenarian. Depending upon leadership and local historical circumstances, different elements were emphasised at different points in time, or were mixed with various results. Anthropological treatment of these movements has varied accordingly, combining cultural, psychological, political, economic, and theological analysis. More recently, commentators have opted for an increasingly self-reflexive and deconstructive focus: cult movements have proven 'good to think', as the gaze is shifted back to the Western fixation with desire and the irregularities and irrationalities of bourgeois society. They have perhaps proven similarly useful, albeit in different ways, for their alleged adherents, who think both with and against them. Yet ultimately, we find that the variety of social movements prevalent within Melanesia continues to highlight Western assumptions underpinning interpretations of history and cultural difference.

Early Encounters

The first recorded sighting of the Lihir Islands was in 1616, by Dutch explorers Jacob Le Maire and William Cornelisz Schouten who observed it from the south at some distance as they hugged the northern shores of New Ireland. The islands were first named Gerrit de Nijs Eylandt by another Dutchman, Abel Janzoon Tasman, in 1643, when he navigated through New Guinea onboard the Heemskerck. He approached within two miles, and an artist onboard named Isaac Gilseman sketched the first image of Lihir (see Sharp 1968: 56).

The first contacts that extended to any form of trade or engagement were probably between Lihirians and whalers from the late 1830s until the early 1880s. Whalers operating in the Bismarck Archipelago during this period frequently sought wood to fuel their onboard processing, as well as food. German naval officers produced the first map of Lihir and the large bay of Luisehafen in 1880. Germany first raised the flag in North East New Guinea in 1884, but it was not until 19 May 1900 that the German administration arrived in Lihir in the form of Governor Rudolf von Benningsen and Prof. Dr Robert Koch, accompanied by the ethnographic collector Lajos Biro. During the 1880s, greater numbers of Lihirians became involved in the labour trade for the Queensland, Fiji and Gazelle Peninsula plantations, in what was often regarded as 'blackbirding' — the kidnapping of unwilling ill-fated islanders. Missionaries were habitually prone to describing these labour vessels as 'slavers', based on the assumption that islanders failed to comprehend the realities of the contract (Scarr 1967: 139). Historians have since reassessed these assumptions, suggesting that, once islanders were aware of the contracts, they used the opportunity to travel and signed on (and off) of 'their own free will' (Firth 1976: 52).

Large numbers of Lihirians signed on as plantation labourers for the Queensland sugar industry.[3] Men and women were recruited from various places, with little uniformity across the region. For instance, in 1883, 649 men from Lihir signed on — an extraordinary number given that the population then was well under 3000 people — compared with only 368 from Tanga, 37 from Feni, 28 from

---

3 Given the comparatively high number of Lihirian men who initially signed on for the labour trade, it is tempting to speculate whether this activity also had cult-like qualities, and whether Lihirians interpreted recruiters as returned ancestors and thought that this avenue might offer a path to 'deliverance'.

Tabar and 240 from mainland New Ireland (Price and Baker 1976: 116). Lihirians also participated in the Fiji labour trade, albeit offering only eight men between 1876 and 1911.[4]

New Irelanders recruited to work on the plantations throughout Kaiser Wilhelmsland (mainland New Guinea) had also come to realise the associated risks, which Firth suggests are best summarised by a single statistic — an average annual death rate of over 40 per cent of the 2802 islanders who passed through Kokopo control station on their way to western Kaiser Wilhelmsland and Astrolabe Bay plantations between 1887 and 1903 (Firth 1976: 53). Before long, New Irelanders were loath to travel and work in these areas, and began responding aggressively towards traders and recruiters. Kaiser Wilhelmsland was gaining a reputation in the New Guinea islands 'of being the place where there was "no kaikai [food], no Sunday, plenty fight, plenty die"' (Firth 1982: 38). Violent interactions between traders, labour recruiters and New Guineans were common in northwestern New Ireland, partly fuelled by community anger and resentment over the loss of kinsmen in plantation labour:

> Men sought vengeance for the loss of kinsmen on plantations far from home. NGC [New Guinea Company] recruiters in the area of the Lihir Islands near New Ireland were subjected to what appeared to be a planned attack in March 1884, when nine new recruits fell upon them with their bare hands while kinsmen gave support with axes. The German helmsmen and at least two New Guinean crew were killed (ibid.).

Reluctance to engage in wage labour stemmed from the atrocious working conditions and a more general realisation that the lot of the labourer was not necessarily better than village life. Such sentiments were captured in the words of one Lihirian who, in 1900, reportedly asked Governor Rudolf von Benningsen 'why villagers should go to the plantations when they had plenty to eat at home' (Firth 1976: 54).

Not all outside engagement was violent. In 1907, Otto Schlaginhaufen left Germany with the Deutsche Marine-Expedition for the Bismark Archipelago. He reached the shores of Lihir at Leo, near Palie, in 1908. During his time in Lihir, he recorded 19 traditional Lihirian songs on wax cylinders, mapped the main island of Aniolam, and documented aspects of Lihirian culture (Schlaginhaufen 1959: 133). The renowned ethnologist Richard Parkinson, who worked throughout New Guinea between 1882 and 1909, also visited Lihir and recorded local cultural traits (Parkinson 1907).

---

4   By comparison, some 36 males from Simberi (Tabar group), 104 from Lavongai (New Hanover) and 325 from mainland New Ireland made their way to Fiji (Seigel 1985: 53).

In the final years of German rule, the annual reports state in an ominously prophetic tone that 'hitherto … the administration has not been able to play a more active role here [in Lihir] because of poor communications' (Sack and Clark 1978: 338). The Australian administration, which followed the Australian occupation of the Territory in 1914, did not bring substantially greater economic development or considerably improved communication with Lihirians, averaging at best one patrol a year until at least the mid-1960s. The Australians inherited from the Germans an organised system of leadership, relying on village 'police' and their assistants, known as *luluais* and *tultuls*,[5] to implement colonial authority (Rowley 1958: 217; Firth 1982: 2) (Plate 3-1). With the cessation of German rule, the first task for the Australians was to 'reassure' people that although the Germans no longer held sway, the hierarchical order remained the same: Whites were vested with power to direct the lives of Melanesians, or as Firth (1982: 2) appositely states, 'it was the old colonial order under new management'.

Between the wars, Lihirian engagement with the cash economy was largely confined to limited copra production and petty trade within the domestic sphere. Reminiscences collected from elderly Lihirians indicate that economic productivity was especially low during these years. Lihirians, conscripted from Lihir and other parts of New Guinea such as Rabaul, participated in World War Two as labourers, carriers and messengers for both Japanese and Allied troops. Lihirians received remuneration for their assistance to the Allied troops, and compensation for war loss and damage, including that resulting from activities by the Japanese.[6] Most oral accounts stress high numbers of males absent from Aniolam during this time and the harsh conditions under the Japanese (Zial 1975). Several older Lihirian males from Aniolam recounted their forced labour constructing airstrips on mainland New Ireland. Many recalled Japanese brutality, describing gruesome deaths for those who relaxed or refused to work. It is likely that post-war reluctance to be involved in plantation labour was shaped by these experiences.

The Pauperisation of Lihir

By the 1950s, economic development was still relatively limited. A single copra plantation had been established on the plateau at Londolovit (now the site of the mine camp and town), employing imported labour from New Ireland, New Britain, Bougainville and mainland New Guinea, particularly the Sepik region. Lihirians were apparently not only unwilling to work under foreigners but

5   The words *luluai* and *tultul* originate from the Kuanua language in East New Britain.
6   For example, Namatanai Patrol Report (PR) 8-1948/49 records paying out 248 pounds and 19 shilling to people in Lihir, mostly for pigs taken by the Japanese.

seemingly uninterested in local production. There were no local copra groves; locally owned plantations were small and used mainly for feeding domestic pigs. While the Catholic mission, which was established at Palie on Aniolam in 1902 (Trompf 1991: 169), ran a small trade store, it did not yet deal in copra. Government administrators saw agricultural potential in Lihir, but their attention was focused elsewhere in the district.

Plate 3-1: Lihir *luluai* in front of a *tlaliem* canoe, circa 1932.

Photograph by E.P. Chinnery, courtesy of the National Archives of Australia, A6510–1478.

All of this provided little incentive for Lihirians to develop their own plantations (Namatanai PR 4-51/52), and they were left by the administration to their own devices and ingenuity. Yet when Lihirians attempted to act upon their desires for (economic) self-improvement, their actions were consistently thwarted by an administration that held Lihirians incapable of running their own affairs, said to be the result of 'inbreeding' that produced a higher ratio of natives with 'sub-normal intelligence' (ibid.). Such apparently below average acumen did not stop Lihirians from raising nearly A£2000 towards their cause through 'donations' for the 'Lihir fund', which they were advised by the administration to deposit into a trust account at the Namatanai Sub-District office.

In 1951, Lihirians attempted to purchase a small work boat, to be Lihirian owned and managed, that would provide more regular travel to Namatanai on mainland New Ireland. Although the Catholic mission ran infrequent (and overcrowded) trips to the mainland, this service was insufficient for any regular copra trade. This early attempt to assume greater control over the direction of local economic activities was swiftly halted. The administration decided not to support this cooperative initiative, concluding that there would be too many logistical problems associated with local ownership and management of a boat. In years to come, the debacle surrounding the wreck and subsequent demise of the locally managed MV Venus on the neighbouring island of Tanga (Foster 1995a: 54–6) confirmed for the Australians the futility of locally owned and managed boats, causing them to rule such local initiatives 'out of the question' for Lihir (Namatanai PR 4-59/60).

By 1952, the administration had made the important observation that Lihirians were gaining a sense of economic self-consciousness:

> Lihir Natives are awakening to the realisation that they are the poorest natives financially in the New Ireland District. The Lihirs have comparatively small holdings of coconuts and no alternative cash crop (Namatanai PR 6-52/53).

Even if the administration had possessed the foresight to act on these observations to avoid future political unrest, they were still constrained by scarce resources. It was some years before Lihirians connected their economic status with a government that denied their equality. Lihirians were beginning to perceive a process of 'pauperisation'. Like Jorgensen's observations among the Telefolmin in the New Guinea highlands, this 'was rooted not so much in the "objective" features of the local economy — people did, in fact have more in the way of cash and material goods than before — as in their sense about where they stood in relation to future shares of worldly goods' (Jorgensen 1981: 66). These early murmurs of discontent would later flourish in the Tutukuvul Isakul Association movement, and find new life in the context of large-scale resource extraction.

Throughout the 1950s, the Lihirian copra industry advanced minimally, and Lihirians remained peripheral to New Ireland economic activity. Both the Huniho and Lakakot Bay plantations (on Aniolam) were established to complement the one already established at Londolovit, but they hardly offered encouraging prices for copra. The estimated total annual income recorded for Lihir in 1959 was A£8355, of which A£4015 was earned by 162 non-Lihirian plantation workers who came from New Ireland, New Britain, Bougainville and the Sepik District. Only 33 Lihirians were employed as labourers, earning A£1720; the remaining money was raised through personal sale of copra and foodstuffs, including pork (Namatanai PR 4-58/59). The fact that few Lihirians chose to work on these plantations amazed and frustrated plantation managers. While negative associations with indentured labour and wartime experiences undoubtedly influenced Lihirian reluctance, their desire for economic activity, tempered by their desire to have this exchange on their own terms, continued to be apparent — and has since been a recurrent theme in the relationship between many Lihirians, the mining company and the State.

The administration regarded Lihirians as 'lazy' and 'economically lethargic' (Namatanai PR 4-59/60), and so concentrated its efforts and resources elsewhere. Although various patrol officers noted Lihirian interest in forming copra societies, and recognised this interest could be harnessed if Cooperative Officers were sent to Lihir and the administration had a more regular local presence, it was decided that, until Lihirians showed more signs of genuine self-activity (more economic progress without external assistance), the administration would continue looking elsewhere in the district. Such policies kept Lihirians entrenched within a cycle of economic marginality.

During the following years, Lihirians were encouraged to establish more individual copra plantations as a basis for local economic development and political advancement. The administration was perturbed by the fact that Lihir showed so much unrealised potential. If Lihirians were to be incorporated into the local government council system, they would need to show more consistent efforts in their copra production. In 1959, Lihirians had over 32 000 mature, and 28 000 immature, coconut plants, which the administration estimated should yield at least 170 tons of copra annually. Instead, Lihirians were only producing between 60 to 70 tons (Namatanai PR 4-59/60). These results did not endear Lihirians to administrators, who believed that it was possible to 'develop the natives' by instilling the value of assiduous work and entrepreneurialism, but regarded those who failed to respond positively (according to Australian standards) as hopeless.

For the Australians, the acknowledged absence of necessary infrastructure for Lihirian economic development, such as a road, vehicle transport, sea ports and regular marine transport did not excuse a lack of industriousness. Lihirians were encouraged to make use of the marketing channels already available through the mission and the three plantations. In 1959, Lihirians were exempt from taxation, presumably due to their low economic output, but in 1960, taxation was reintroduced as an incentive to economic activity, with explicit instructions that no exceptions would be granted where people had the opportunity to make copra. This still failed to lift production levels, and the administration argued that this was the result of a peculiar attitude among Lihirians, where people only did enough work to cover their annual A£1 for tax and A£1 for clothing (Namatanai PR 10-61/62).

Admitting to years of neglect, the administration argued that the only remedy for Lihirian economic stagnation was increased contact with Europeans, in the belief that this would somehow develop the necessary economic sensibilities. At the same time, even as enthusiasm for cooperative societies increased, indicating that not all Lihirians were as 'backward' as the patrol officers liked to imagine, any economic advancement was still inhibited by the lack of qualified staff and appropriate transport, and for the immediate future, there was no intention to rectify this situation (Namatanai PR 13-64/65). Interestingly, the patrol reports repeatedly state that Lihirians had not lost 'faith' in the administration despite the lack of attention, and that Lihirians were still optimistic about what the administration might do for them. This might be interpreted simply as an administrative delusion, or an indication of government arrogance, ignorance and conceit. Alternatively, it may partly explain why support for the administration later declined rapidly in favour of emerging local prophetic leaders who promised more immediate returns on local investments. New expectations fostered by local leaders who preached radical messages of change would have lasting effects on Lihirians as they began to articulate their desire for greater control over their economic and political future.

'Improvements' and Approaching Impairments

Between 1965 and 1970, Lihir underwent three significant developments that paved the way for the TIA (which initially began on New Hanover). These were the establishment of 'progress' or 'cooperative' societies, incorporation into the Namatanai Local Government Council (NLGC), and the emergence of entrepreneurial 'big-men',[7] who gained new status through their monopoly on copra production. The administration initially took these events as signs of

---

7   There has been considerable debate in the Melanesian anthropological literature over the term 'big-man' that denotes a particular style of Melanesian leadership. Although Melanesian 'big-men' have been contrasted to Polynesian 'chiefs', many anthropologists have contested the use of this term for all Melanesian leaders (see Sahlins 1963; Godelier and Strathern 1991).

progress that validated its own policy; but in reality, these were less the result of 'obedience' to administrative directives and more the cumulative result of nearly 70 years of sporadic engagement with outside influences.

The first of these changes took place after a visit in 1965 from Nicholas Brokam, the Member for New Ireland in the first House of Assembly for PNG.[8] He came to gauge interest for a locally owned copra boat, believing it would give Lihirians a greater sense of connection to the mainland. Sigial, a *luluai* from Lesel village (on Aniolam), took it upon himself to form a committee with other big-men, with the intention to raise funds through a form of taxation and publicly discuss the need for a government school, increased copra plantings and something described as *lo bilong Brokam* (Brokam's law). The administration found this an 'alarming situation', particularly as there was no consistency in the 'tax' levied; women were only occasionally exempted, and men were expected to pay between A£5 and A£10 (Namatanai PR 15-65/66). A man of immense ambition, Sigial was partly acting on faith in the administration, and partly attempting to make himself a man of consequence — describing himself as the 'Member for Lihir' — who could deliver economic prosperity if people abided by the 'laws' he delineated as the key to success. The administration was wary of the potential for disturbance as a result of Sigial's activities, but concluded that, as no cults had 'broken out', all they could do was keep watch on his movements (to the extent that this was possible through annual visits).

This was the first recorded instance of Lihirians attempting to transform their society by mimicking the salient features of administration practice. While ritualistic performance of 'official' activities was a common feature of numerous movements throughout Melanesia, this was not mere copying or blatant irrationality. In Lihir, this was an attempt to harness the 'power' and 'success' seemingly inherent in the administration's activities, or what Douglas Dalton (2000: 290) might describe as 'corporally doing power'. In the years to follow, and into the mining era, finding the right 'formula' or practices became central to Lihirian attempts to understand and control modern processes of production and wealth accumulation; and in each instance, this was inspired by an imagined future that repositioned Lihir itself.

Sigial's activities coincided with the formation of a number of small cooperative societies, some of which were probably spurred on by his messages. These represented some level of economic progress. On the outer islands, people began

---

8   Brokam was the Member for New Ireland from 1964 to 1966. Throughout 1965 and 1966, he made several enquiries in the House of Assembly about obtaining work boats to better serve patrol officers and communities in Namatanai District, though as we shall see, these requests were not fulfiled on account of limited government resources. He recognised the correlation between discontent in New Hanover and limited development opportunities and government assistance, and argued that a similar situation would arise in other isolated island areas if assistance was not provided.

forming themselves into functioning societies that worked together to market copra. Similar activities occurred on Aniolam, as people formed *wok sosaiti bilong tesin* (plantation work societies). People paid membership fees to join, and spent two days a week maintaining plantation groves 'owned' by various members. The administration understood these activities as confirmation of its modernist ideals:

> Community effort will teach them the basics of working together as a unit and will place the progress of Lihir as a whole uppermost in their minds, and not the fostering of tribal differences, a stranglehold on progress at the best of times. These later will no doubt die a hard death, having been the accepted way of life for centuries, but their exit will be hastened when the benefits of community effort are made manifest. A square mile of healthy young palms speaks so much louder than an acre of ancient ill kept ones (Namatanai PR 15-65/66).

Copra production levels increased but never matched predictions. The rise was partly due to the establishment of new cooperative societies, but also the emergence of entrepreneurs who began to alter Lihirian economic, political and social conditions. These enterprising big-men began drawing on introduced forms of governance and economic activity to reinforce authority within existing frameworks of power. Both the 'traditional' and the 'modern' were integrated: *bisnis* (petty trade in the form of the copra industry) was in no way extraneous to or practically insulated from the matriline or the realm of *kastom*.

Bell's (1947) early descriptions of Tangan exchange correlate with anecdotal accounts from Lihir and provide insights into the ways in which New Ireland big-men managed and utilised introduced items and money within their local political economy. These accounts suggest that Lihirian feasting and exchange became grander and more competitive as big-men controlled and exploited new forms of wealth gained through plantation wage labour, either by themselves or through those under their influence. Notwithstanding the fact that exchange practices were radically altered through 'pacification' and the introduction of steel tools and new forms of wealth, the changes that occurred as new entrepreneurial big-men gained a stronghold on the nascent copra industry in the late 1960s and early 1970s represented a new era in Lihirian economic relations (for examples elsewhere in PNG, see Strathern 1971b; Gregory 1982; Clark 2000). Their ability to manage clan and village labour, which assisted their domination of the cash economy, also bolstered their local authority. This did not create a rural proletariat, but it did mean that the majority of Lihirians were subordinate to the success of these men and remained peripheral to 'commoditisation'. As in other Melanesian societies engaged in cash cropping, the colonial economy began to be reconfigured in terms of big-man politics (see Finney 1973; Strathern 1982b; Foster 1995a). These men established small trade stores, maintained large copra groves that employed relatives and village members as casual labourers, formed 'progress societies', and provided transport services with motorised canoes. Leadership was consolidated around their management of a local political economy that was completely enmeshed with council activities and the control of local copra production, placing them among the 'biggest' men in Lihir. Their strategic effort to subsume the cash economy into the local exchange system represents an early attempt on the part of Lihirian people to control their entry into modernity and global capitalist exchange.

In 1967, Lihir was officially incorporated into the Namatanai Local Government Council (NLGC) and, within one year, the council was fully entrenched in Lihir with the appointment of local councillors. Elderly *luluai* and *tultul* were replaced by younger and more educated men. While some *luluai* stayed on as councillors (the new local government leaders under the NLGC), many were happy to relinquish their positions, especially those who lacked authority and were unable to carry out their council duties. Despite continual references to Lihirian ignorance and uninterest in political progress, in 1970 it was noted that Lihirians had a 'grossly excessive representation' in the NLGC with their 11 members, far exceeding the number from any other census division on a per capita basis (Namatanai PR 12-69/70). Regardless of whether Lihirians were able to comprehend the Westminster system, or who represented them in the national political arena, their entry into the new government system did not dampen enthusiasm for government-funded development. Indeed, it was this increasing expectation of government assistance and attention — never to be realised — that bred disillusionment, hostility and dreams for an inverted order.

Lihirians were becoming increasingly aware of their relative deprivation in comparison with Europeans and their close neighbours in the region. Reluctance to work for Europeans and non-Lihirians on local plantations reflected the minimal wages and returns offered in this exchange and the inferiority which Lihirians were made to feel at the hands of *ol kiap* (government patrol officers) and plantation owners. Incorporation into the NLGC, which coincided with the formation of cooperative societies and an increased interest in local economic production, fuelled expectations for the changes that the administration promised would arise through 'self-activity'. Inclusion into the new government system brought increased control into their lives, but did not secure greater economic activity in spite of the efforts of some Lihirians. Copra societies might have represented local enthusiasm for collective activity, but without the necessary transport and infrastructural support from the administration, these efforts were bound to fail.

The rise of entrepreneurial big-men, who gained prestige across various realms, introduced Lihirians to unprecedented forms of economic stratification. As younger big-men merged their monopoly over copra production with their local political aspirations, some Lihirians expressed discontent over individual wealth accumulation that was seen as possible only at the expense of others. Combined with resentment towards the existing colonial hierarchy, Lihirians, especially older big-men with waning political influence, were receptive to the radical messages that would arrive from New Hanover.

The 'Johnson' Influence

In 1964, the Territory of Papua and New Guinea held the first national elections to provide representatives for the House of Assembly. In New Hanover, northwest of mainland New Ireland, which had a population of approximately 7000, nearly half the adults refused to follow the prescribed voting method, instead voting for America's President Johnson. These stirrings of discontent soon came to the attention of Lihirians, as people responded with increased antipathy towards the government's inability to fulfil local desires for economic progress and moral equality. According to Dorothy Billings, people in New Hanover had been impressed by US army surveyors working in the district, whose generosity with food, goods and payment for locally hired labour was unprecedented. This became known in New Hanover as 'the American way' (Billings 1969: 13). Frustrated with their marginal status and comparatively slow economic progress under the Australian administration, people from New Hanover voted for Johnson and raised A$1000 to pay his fare to New Hanover. The administration responded to this action with more regular patrols, political education, and violent punitive expeditions. In return, the people of New Hanover refused to pay their taxes, and defaulters were consequently imprisoned (Billings 1969; Miskaram 1985).

The movement was labelled 'The Johnson Cult' (at least among its critics), although it was never a cult in any common sense of the term: it spawned no prophetic leaders, and the emphasis was practical not ritual. There were no visitations, deities, spirit mediums, elaborate doctrines, or epiphanic revelations. Whether the ancestors knew how to produce cargo was unclear and largely irrelevant: clearly, the Americans knew how, and they could teach the people of New Hanover, while the Australians apparently refused to share their *save* (knowledge). The main aim of the movement, which gained several thousand supporters in southern New Hanover and parts of New Ireland, was not political independence but the replacement of Australians with Americans as a means of improving the people's welfare and status (May 2001: 59). In an effort to channel these energies into more productive ends, Father Miller, an American Catholic priest working on New Hanover, encouraged people to form themselves into an 'investment society', which they called the Tutukuvul Isakul Association (TIA).

Miller's American heritage proved influential, often reinforcing the conviction among adherents that the Americans would somehow come to their aid. The TIA gained momentum throughout New Hanover, with strong support from Johnson 'cultists'. Although TIA membership required payment of government taxes, many members remained resistant towards the council and refused to pay.

Lihirian Adoption

The movement soon spread to Kavieng District, and groups organised themselves under the new title Tuk Kuvul Aisok (TKA), partly in an attempt to separate themselves from the stigma of the 'Johnson cult'. The TKA was reportedly introduced to Lihir in 1969 by Theodore Arau from Matakues village on Aniolam. Arau was described by Yngvar Ramstad as a quiet and shy man, about 45 years old at the time (Ramstad n.d. 1).[9] Arau was never a big-man in the traditional sense, but he had once been a catechist for the Catholic mission and then a native *doktaboi* (doctor boy, or medical orderly). While living in Kavieng, Arau became a member of the TKA movement, later encouraging other Lihirians to join. Within three months, close to half the adult Lihirian population of approximately 4500 people had reportedly joined the association, and within a few years, as its influence and membership increased, every village was split between members and non-members. Inspired by the activities in New Hanover, Lihirian members also refused to pay taxes. Initially, the administration took a hardline stance towards the TKA, jailing those who refused to comply with territory laws, assuming that this would deter others. As was the case in New Hanover, however, this only made martyrs of the people who were prosecuted, strengthening local resolve and resistance.

At various times, the TKA resembled a 'progress society' or 'self-help movement', emphasising pragmatism over religion and self-help over myth. The TKA held regular working days that often drew people away from council tasks such as cleaning villages, maintaining paths, or building roads and bridges. Efforts were supposedly concentrated on clearing and maintaining TKA copra plantations in various villages. The divide between government and TKA work caused obvious rifts but, according to several elderly (pro-administration) Lihirians, the problem was not so much the division of labour, but rather that TKA workdays were often spent holding meetings and discussing possible dates on which the cargo would arrive. The TKA reportedly charged its members A40 cents for every day they did not spend cleaning and maintaining TKA coconut plantations, which may account for the reluctance of some entrepreneurial big-

---

9   Ramstad conducted ethnographic fieldwork in Lihir in the late 1960s for his doctoral thesis at the Australian National University. He did not complete his thesis in English (although he may have done so in Norwegian). He wrote three short seminar papers on Lihirian society, focusing on kinship, ritual, and the TKA movement (see Ramstad n.d. 1, 2, 3).

men to join the association. Membership fees ranged from A$2 for women to A$10 for men. Most people were unsure where their money was being stored or for what purposes it would be used, except for a sure belief that they would soon see a return on their 'investment'.

In 1969, Patrol Officer D.M. Donovan conducted a 'slow and thorough' investigation of all 'cult activities' in Lihir (Namatanai PR 8-69/70). The 'outbreak' marked the beginning of more concentrated administrative attention, yet this only strengthened TKA support and local resentment towards the administration which had supposedly denied them knowledge and blocked their 'road' to progress (Burridge 1995: 184). Within a matter of years, administration officials again admitted that for too long they had concentrated their efforts on Tanga, Anir and Namatanai, and that it was 'now Lihir's turn' (Namatanai PR 8-69/70). But their ineffectual reparation only consisted of more admonishment and rational modernist discourse.

After Donovan's initial foray, it was clear to the administration it was dealing with a 'full-blown cult'. This was confirmed by the high level of membership and the commitment displayed by leading prophetic figures such as Arau (the president), Tienmua (vice-president or 'number 2' for Lihir), and other office-holders such as Kondiak (the 'board member' for Lesel village) and Pesus (the 'clerk' from Londolovit village). The TKA's ideologies were closely aligned with those on New Hanover, but if Billings is correct and the TIA was more practically orientated than ritualistic, the Lihir version represented a significant departure.

Arau's prophetic messages and 'rules', which he supposedly received from New Hanover, centred around three dominant themes: the need to resist the council (and the administration); the prediction that change was imminent and likely to come from America; and the belief that Lihirian ancestors were thoroughly implicated in this process. While Donovan was carrying out his investigations, he collected a list of incidents and 'rules' which he translated from Tok Pisin into English. These summarise TKA beliefs and provide some insight into local activities at the time:

Teachings of Arau and Tienmua

1. Now is the time to get rid of the Council.

2. If you see a 'Kiap' coming, get your basket and run away into the bush.

3. If you are taken to court, TKA will bring you back.

4. If a person is not a member of TKA, he will be a 'rubbish' man and no money will be forthcoming to him. Also if he wants to sell one or two bags of copra to any member of the TKA, he will only receive 50 cents for it and be told to go.

5. All plantations which we have planted are for temporary measures only. If the 'Egg' hatches, they will be destroyed or left to feed to pigs.

6. If the 'Egg hatches' you will not have to work to find money. You can rest but money will come to you like flowing water.

7. If a person said something bad against TKA that person will become known to the TKA automatically.

8. TKA is a country we have not seen.

9. America is one of the true country and one is at Palie, Father J. Gliexner.

10. USA is one of the Countries that will never die and some of them are here — Fr. Tom Keller (Namatanai); Fr. Miller (Lavongai); Fr. David Milmila (Duke of York); Fr. Peter Vavro (Tanga). These priests are from USA, the country that never dies, but lives for ever.

11. And Fr. P Vavro is from Mazuz (Lamboar Lihir Island), he is not from USA.

12. Now a big ship and an aeroplane are being loaded with cargoes. Both are not fully loaded yet, but they will be sent to us when they are ready.

13. A man and a woman whose wife or husband has died shall not re-marry. They shall await the arrival of his wife or her husband when the 'Time Change'.

14. Those children who attending schools today can be saved, but we will await the time when life changes and then knowledge will come unlimited.

15. There are two types of 'Crabs'. One type went ashore on Siar and the other on Kavin (Kavieng).

16. If a person dies, do not say he is dead, you must say 'He has gone'.

17. If we hear the spirits of the dead, we must not refer to them as 'Ghosts', we refer to them as 'Brothers'.

18. If we call them ghosts or devil, this will stop the arrival of cargoes, time will not change and the 'Road' will not open to us.

19. Supposing this country wants to declare war with another country, our country will destroy that country.

20. It is the same with money, where the white man have changed their face value e.g. 10 cents [for] $1.00 and 20 cents [for] $2.00.

21. All of our money have been sent to the Bishop in Kavieng who will convert into American currency and then send them to America.

22. During March 1970, shipment of cargo will arrive from America. Note: Previous to this THE target date January and Feb were also marked for such.

23. Tienmua and Kaiprot of Konogogo (W.C. Namatanai) have been going around showing pictures from a book to people.

24. Arau and Tienmua have been telling the people that they propose to go to another place by secret means.

25. The people believed that they went to Lavongai in a submarine.

26. All male persons must marry before they could become a member. Because when the time changes, they must go inside the 'House' with their wives.

27. Iaspot [Yaspot] of Malie Island asked Cr. Lusom why the Council knew more about this work. Why haven't they bothered to find out the truth of government laws which are 'eating' the people.

28. On the 5th of January 1970, Toron of Malie Island Committee for TKA collected fees from members at Malie Island. $1.00 for males. 50 cents for females. This they said is to purchase a car for the members use (Namatanai PR 8-69/70).

This list provides one of the few detailed accounts of TKA beliefs during the early stages of the Lihirian movement. Although many of these points resonate with more general themes from New Hanover, some require clarification. This list also provides an opportunity to consider some of the cult activities and enduring TKA beliefs that shaped Lihirian expectations for development and the current political and economic environment.

Arau often used the metaphor of an egg hatching (*kiau bai bruk*) to describe impending change and the delivery of cargo — the new millennium. This was not his own metaphor; it was in use elsewhere some years before the introduction of the TKA to Lihir. Billings recalls that, during the elections in Kavieng in 1964, she heard rumours of 'a Big Egg which was said to be believed to be hovering over New Britain and about to hatch cargo' (2002: 29). Lihirians and other New Irelanders have a history of cultural borrowing from New Britain, and it is quite possible that such metaphorical images were imported from similar movements occurring in New Britain at the time (see Counts 1971). Indeed, many Lihirians were willing to believe Arau's prophecies because of the magical knowledge which he obtained from the Buai cult, which originated in New Britain and the southeastern part of mainland New Ireland. The idea that money would flow like water, regardless of labour input, was also expressed in the idiom of 'live money' and 'dead money' — an idiom which some older Lihirian men later applied to the different forms of income derived from mining. In these terms, royalties and compensation payments are 'live money', wealth obtained without physical or moral diminution, as opposed to the 'dead money' earned through the back-breaking and humiliating work of *ol kagoboi* (cargo boys), cash cropping for minimal returns, or low-paid unskilled work in the mine.

Ancestral Connections to the Last Country

Phrases like 'Time Change' and 'TKA is a country we have not seen yet' represent the genesis of later millenarian concepts expressed in the idea of Lihir becoming a 'city' and the return of deceased ancestors. Arau's prophesies generally appeared vague and non-committal; instead of concrete images, there was just the expectation of some form of revolutionary departure from a condition of subjugation and material poverty. It is possible that, in recent years, Lihirians have retrospectively applied the concept of a 'city' to Arau's prophesies. Certainly, contemporary urban ambitions that draw from images of New York, Singapore and Sydney — now available through increased media access — are vastly different from anything people would have known about in the 1970s, when the weatherboard buildings in Kavieng town were considered metonyms of Melanesian modernity. Alternatively, the first 'city' that some older Lihirians may have seen or heard about was the US air force base on Emirau Island, north of New Hanover, seemingly built in a matter of days or weeks during World War Two, and literally brimming with technology and people. It is likely that the reference to 'crabs' travelling to Kavieng and Siar (at the northwestern and southeastern ends of the New Ireland mainland) is symbolic of Lihir-centred change. Given the insularity of Lihir in this period, these two places possibly represent the 'limits' of the known world for many Lihirians. The vision of becoming a 'city' reflects local conceptions about centres of power, and a continuing concern with Lihir's peripheral status within New Ireland.

Arau's prophesies were expressed in the Lihirian concept of *a peketon*, which refers to waves crashing on the shore, washing flotsam and jetsam onto the beach and then, with the receding tide, carrying the debris to other places: as change (or cargo) comes to Lihir, it will then emanate outwards from the new centre. Martha Macintyre (personal communication, June 2008) said that she asked about the meaning of this term during her initial visits to Lihir in the mid-1990s. Several Lihirians responded that it was a 'white man's word'. They thought that she was 'testing' them to see if they were 'ready' to receive the cargo; over the years, various Lihirians have attempted to prove that they were 'ready' through different activities and quests. Years later, it became apparent that they were referring to the Ancient Greek word *eschaton*, most likely introduced by the Catholic mission, which sounds quite similar to *a peketon* when pronounced in some New Ireland dialects. The linguistic and descriptive fit developed an influential metaphor for change arriving from the outside world: that Lihir would become the 'last country', or the 'new heaven'. While Lihirian millenarian concepts have undergone transformation as new leaders emerged, together with new political goals and the economic changes brought about by mining, the metaphorical concept of *a peketon* has remained central.

In the 1970s, TKA members came to believe that America would eventually replace Australia as the governing body. This might not equate to secessionism, but it does reveal their dissatisfaction with the Australians. The administration naively regarded this belief as a gross manifestation of big-man politics:

> In their social structure, if their head man does not give them what they want, they merely keep replacing the head man until they find one that satisfies them. In the same way they wish to replace New Guinea's head man (Australia) and obtain a new one (America), which in their opinion will be more beneficial to them (Namatanai PR 4-70/71).

It is probable that Lihirian engagement with Americans during the Second World War encouraged people's desire to be annexed to America and their belief that America is the 'true' country that would deliver the promised cargo. This opinion may have been influenced by stories or experiences on Emirau, where US wealth, power and organisation were demonstrated. This confirmed the TIA's strong pro-American messages, and the stories of harmful race relations in New Hanover most likely resonated with Lihirian experiences with the colonial administration. Lihirians deeply admired the people of New Hanover, who had evidently surmounted government persecution and opposition. Letters of encouragement were sent from New Hanover urging people not to support the administration and predicting that New Hanover and New Ireland would soon be a state of America (Namatanai PR 17-70/71).

What Did They Really Want?

Billings argues that, as the descendent organisation of the Johnson movement, the TIA was principally concerned with social equality and not just with material wealth. Being primarily a politico-economic movement, it was not 'fundamentally religious' nor was it 'a manifestation of gross materialism':

> They wanted to understand history, historical forces, and power well enough to maintain their place and their ability to control their lives. Those with the broader view did not like being considered ignorant and poor, and did not like feeling constantly humiliated by white people and by their own educated compatriots. They wanted equal status, and equal knowledge, in the modern world (Billings 2002: 165–6).

Likewise, Lihirians were not merely vulgar materialists. As various commentators have noted of similar movements, such as the Yali cult in Southern Madang (Lawrence 1964), the Mambu cult in Manam (Burridge 1995), the Kaun movement in Karavar (Errington 1974), or the Kaliai cult in New Britain (Lattas 1998), fancy Western stuff is only half the story. Obtaining wealth, goods, and even 'organisation' or 'order' is premised on gaining respect and a sense of equality. Consequently, indigenous questions about the origins of wealth have often been interpreted as questions about relations between Blacks and Whites — questions not only about material deprivation, but about the denial of equal humanity. Through obtaining the desired cargo, Lihirians not only sought to enhance their daily lives, but also to gain the respect of Europeans by possessing what the latter so obviously valued. Through their eyes, we thus begin to see the practical and symbolic qualities of goods, the social uses to which they can be put, and the idea that things are valued not only for their material uses, but because they can be used in social transactions that establish mutuality and respect (Appadurai 1986; Sahlins 2005b).

In the early 1970s, rumours circulated as far afield as Kavieng and New Hanover that Lihirians were erecting large storage houses in anticipation of cargo ships sent by their ancestors. Some Lihirians admitted that these and other structures were attempts to prove that they were 'ready' to receive the cargo. This theme later resurfaced in 2000, when John Yaspot from Malie constructed a large two-storey men's house which many thought resembled a 'hotel' rather than a men's house. Criticism was largely directed at the deviation from traditional style. His rationale for this design was twofold. He thought that such a house might be more appealing to younger men, which was an indication of the declining importance of Lihirian men's houses, and perhaps more significantly, that this would demonstrate that he was 'worthy' to receive the anticipated cargo from his ancestors.

People's preoccupation with cooperative activities reflected a similar concern with 'proving' that they could organise and work as efficiently as Westerners. Lihirians shifted between the practical and symbolic understanding of work as they attempted to understand Western success and power. However, given that Lihirians wanted America to replace Australia as the *papa kantri* (father country), much of this was designed to impress both the ancestors and the Americans, not the Australians. For some, America was conflated with notions of a benevolent father, the home of the ancestors and the fulfilment of TKA prophecies. In this instance, the enduring relationship between Lihirian ancestors, Whites, and receiving cargo (or obtaining development) is located in people's ability to prove their 'modernity' or their 'worth'. This contrasts with Leavitt's (2000) examination of cults among the Bumbita Arapesh of East

Sepik Province, where people aimed to fix the rift in their relationship with ancestors who were supposedly withholding the cargo until their relationship was reconciled.

However, even though Lihirian political and economic ambitions were deeply rooted in the desire for moral parity, arguments commonly employed by anthropologists attempting to make sense of Melanesian politico-religious movements only partly capture the reality of contemporary Lihirian ambition. As we shall see, by the time mining operations had commenced, Lihirians no longer articulated their aspirations only in terms of equality. They wanted unlimited access to the 'blessings' of industrial technology, but also control of these processes, ensuring their autonomy in the distribution of wealth, while expatriate mining personnel would ultimately become subordinate to Lihirian management.

Divisions and New Directions

Not all Lihirians were convinced that Arau was capable of delivering on his promises. As the association expanded and consolidated its membership, tensions increased between members and non-members, as people disagreed over which 'road' would lead to the desired destination of modernisation. Some sided with the administration out of loyalty and the belief that their own (relatively) close connection with *ol kiap* and other Whites would eventually bring them closer to what they sought. Similarly, councillors rarely joined because the administration provided them with political authority and a consistent (albeit meagre) income. John Yaspot's assertion that the councillors were colluding with the administration and deceiving the people is indicative of the tension between councillors and TKA members.[10] Disagreement over the correct 'road' to achieve dreams of emancipation has continued to characterise political events during mining activities and negotiations about development (compensation, royalties, infrastructure and business). Persistent tensions between the desirable and the possible, and between 'truth' and 'deception', have been central to Lihirian people's relationships, both among themselves and with Whites, governing bodies, and other external agents.

In 1973, a young and educated member, Bruno Sasimua, aided by Father Tom Burns from the Catholic mission station at Palie, attempted to transform the TKA into a registered business group called the Tutorme Farmers Association (TFA).[11] The administration commended this effort, recognising it as 'an honest

---

10 Yaspot was a vocal advocate of TKA on Malie Island and showed formidable opposition to councillors and the administration. In a twist of irony, Ambrose Silul, his first-born son, later become the president of the Lihir Local-Level Government.

11 *Tutorme* is another word that can be translated as 'we stand together'.

attempt to create business opportunities amongst its followers' (Namatanai PR 16-72/73), and promised to send more regular boats to ship local copra. The promise was not kept, and as national independence approached, Lihirian attention riveted on the new national government that was expected to 'deliver the goods'. The death of Arau in 1975, the year of Independence, temporarily threw the association into chaos. Lihirians supported Independence, which they expected would bring the desired change, but their expectations were simply beyond the capacity of the new national government. Ferdinand Samare assumed the TFA presidency in 1977, and anti-government sentiments flourished as TFA members refused to pay taxes or vote for anyone but Jesus Christ. Samare and others were jailed, which again only made them martyrs in the eyes of their supporters. While the association had always uneasily existed as a combination of 'development association' and 'cargo cult', during the latter part of the 1970s, there was a marked withdrawal from both 'politics' and 'business'. Frustration over the lack of change delivered by the national government led TFA members to seek more esoteric options. Filer and Jackson (1989: 174) noted that visions, prophesies and supposed miracles sustained people's belief in the new millennium, and members met in designated areas, where they prayed, sang and performed dances.

The Nimamar Association

The association gradually shifted away from pragmatic endeavours, abandoning its links with New Hanover and its nominal ties with 'farming'. There was a greater push towards self-government, and TFA leaders re-formed under the new name of Nimamar — a sort of acronym derived from the names of four islands in the Lihir group that is meant to imply their unity.[12] This represented an official reformation of the TIA and TKA movements, as this group would eventually turn into the Nimamar Rural Local-Level Government.

Nimamar hardly enjoyed unanimous support, yet Lihirians were clearly united in their Christian faith, regardless of denominational affiliation, and more importantly in their antipathy towards the State, which became abundantly clear in the 1980s as Lihirians entered into negotiations with the State and Kennecott. There was a stated withdrawal from 'politics' and 'business', at least in the sense in which government officials would understand these domains. This shift was compounded by the belief that the prospective mining project

---

12   The name of the main island of Lihir, Aniolam, is derived from the vernacular terms *anio*, which means 'land', and *lam* which means 'big'. Due to dialect differences, some Lihirians drop the A from Aniolam, and simply say or write Niolam. The acronym, Nimamar, which starts with the first two letters of Niolam, follows this trend.

was the fulfilment of Arau's earlier prophesies. Many Lihirians were suspicious of the government's interest in the project, believing that the gold was theirs and that the State might ultimately benefit at their expense.

Plate 3-2: Former Tuk Kuvul Aisok meeting area in Matakues now used by church groups for prayer meetings.

Photograph by the author.

Nimamar leaders preached a form of millenarianism combined with some singular economic and political objectives. The desire for a radical re-ordering of society was made explicit in a letter which Ferdinand Samare wrote to Filer and Jackson in 1985. There is an element of continuity with the list recorded by Patrol Officer Donovan, but also a strong political and economic slant that reflects the failed expectations of the post-Independence period and the anticipated onset of the mining era.

Nimamar Association, Lihir Island

In 1968 Theodore Arau brought word of this association to Lihir Island. At that time it was known as the TKA (Tutukuvul Kapkapis

Associaton).[13] This title means "stand together and work the land". It was a basic principle of this association that when a man wanted to join he had to pay a membership fee of K12. After paying the K12, he then had to answer three (3) questions:

1. Do you believe in God? (YES)

2. Do you believe in the work of the Association? (YES)

3. Are you sure that your beliefs are proof against temptation? (YES)

When these three questions had been answered, Arau himself, who was then the President of this association, would explain some fundamental points to the new member.

The following are some of the beliefs of the Association, in other words the basic points outlined to new members. When a man becomes a member, he joins his father or mother or any other close relative who is already dead. One day, these dead parents or relatives are bound to return and rejoin us. One day our lives will improve. We are bound to receive all the various things presently possessed by Europeans. Right now you may be short of money, but one day there is bound to be enough money for all, and lots of other things besides. There will be nice houses to live in and life generally will be wonderful.

The following are the most fundamental points that Theodore Arau used to make. As long as a man's faith remains unshaken, he will observe the fruits of the Association's work as follows:

1. This is not really an association, it is the final country.

2. One day our parents, uncles, aunts, grandparents and other relatives are bound to return, to shake hands with us and eat with us.

3. Now you are poor, but one day there will be so much money around that it will be like rubbish.

4. Now it is the capitalists who have all the money and commodities, but when the dead return to life, things will be different: the capitalists will become poor, and the poor man will become rich.

5. At some point, once our work has got started, we shall join the CHURCH. (this has since happened, and now we have joined the Church:

   • Custom has become part of our worship.

---

13   Although various terms were employed throughout northern New Ireland for the acroynyms TIA and TKA, all referred to the central theme of 'working together'.

- [We] bring out the cross and erect it in the village.
- [We] vote for the name of Christ.
- We shall hold onto our bibles.
- A visitor will come and (1) a statue of the Virgin of Mary will travel around and (2) the Pope will travel around in PNG. (This idea originated with the Association.)
- Some men will come and obtain information [!].
- Some men will come and investigate our work [!].
- The Company is bound to appear at Landolam, and it will have four (4) main bases: (1) Landolam; (2) Londolovit; (3) Lamboar; (4) Wurtol.
- Many of the things stated or written above are products of the WORLD COURSE.)

The Company has now appeared and is based at Landolam (Putput). The Association is not surprised by this. We in the Association knew ages ago that this would happen, at least from the time that Arau introduced the Association in 1968. That was 15 years before [the Company arrived].

When the Company was due to appear and prospect for gold, there were bound to be foreigners arriving to do the work. The Company has already appeared at Landolam. We knew about it in advance, because Theodore Arau had already told us its title (FINAL COUNTRY). BHP (or BISP?) has yet to appear.[14] If BHP (or BISP?) appear, they will immediately teach us how to produce all sorts of things. This will be the start of the FINAL COURSE. This is also known as the USA. But it is not the American USA, it is the beliefs of the Association.

This USA will produce the Final Country. Once all these things have come to pass, Lihir will command the whole Country. This in turn will include the whole World.

N.B. This is the FINAL KNOWLEDGE, and it is the knowledge of the Final Country. Once we got into the Bible, we discovered that it contained the same beliefs as those which are the strength of the Association. So we in the Association have been strengthened and so have our beliefs. Having studied the Bible, we have come to believe that the Church is the road and the key. Jesus himself originally gave the Church the key to the Final Country. Jesus himself is the head of this FINAL COUNTRY.

---

14 Filer and Jackson assumed that 'BHP' stood for the mining company of that name (Broken Hill Proprietary Ltd), but could not decipher the reference of the initials BISP.

N.B. All this is not a dream or a manifesto. It is a form of knowledge or belief, but men still have to work at it. God put men on this earth, and men must work to follow his plan, to develop this earth in accordance with our Father's own will. To accord with the will of the God the Father, God himself has a plan. It is the TEN COMMANDMENTS. If men are going to change this earth and renew it, Jesus alone is the road and the law by which it can be done:

1. Love God above all things.

2. Love your brother as you love yourself.

HERE WE MAKE KNOWN A NUMBER OF IDEAS OR STATEMENTS, BUT THEY ALL COME FROM ONE BASIC BELIEF. THE ASSOCIATION BELIEVES THAT THE DEAD WILL RETURN TO LIFE. THE BIBLE STRENGTHENS US IN THIS BELIEF. THROUGHOUT THE HISTORY OF THE TKA, [which then became] THE TSA, TFA, TIA, AND NOW NIMAMAR, THIS BELIEF HAS NOT CHANGED. ONLY THE NAME HAS CHANGED, WHILE THE BELIEF REMAINS THE SAME.

Nimamar Association,

Lihir Island,

New Ireland Province,

PNG, Last Country.

HERE ARE SOME FURTHER POINTS ON WHICH THE NIMAMAR ASSOCIATION IS UNITED:

The Final Course:

*   THE NATURE OF MONEY WILL CHANGE.

*   THE STATE WILL BE ABOLISHED.

*   THE ASSOCIATION WILL BECOME THE GOVERNMENT.

*   LIHIR WILL BECOME A CITY.

*   SCHOOLS WILL BE ABOLISHED.

*   THERE WILL BE UNIVERSAL LITERACY.

*   THERE WILL BE TWO CLASSES OF PEOPLE:

*   "LIFE FOREVER";

*   "SUMMON PRIZE" (Filer and Jackson 1989: 367–9).[15]

---

15   Filer and Jackson thought that the phrase 'summon prize' referred to some form of 'hard labour'.

This manifesto, which Nimamar members claimed was not a manifesto, was supported by a second letter given to Filer and Jackson by the Putput 2 branch of the Nimamar Association on 26 November 1985. This letter, which directly addressed the impending mining activities, reflecting the village's proximity to the exploration camp, expressed concerns about migrants coming to work or to establish businesses in Lihir, about the need for Lihirians to be in control of new economic developments, and criticised the government for ignoring Lihirians.

Filer and Jackson (1986: 120) suggest that Lihirian hostility towards formal political institutions largely stemmed from their experience of government and business, and the belief that these institutions had proven to be a hindrance in unifying the community. Although Lihirians had always been politically divided between the interests of different clans and lineages and their respective leaders, the external pressures of the twentieth century prompted a desire for a new form of unity. Consequently, Filer and Jackson settled upon the term 'ritual communism' (ibid.: 116) to describe the mixture of religious commitment, aspirations for economic progress, or a greater share of worldly goods, hostility towards the State, and a desire for increased social and political unity. These themes have strongly persisted throughout the period of mining, manifest in new forms and movements, but ultimately directed towards similar outcomes.

Having been elected to the New Ireland Provincial Assembly in the 1986 provincial elections, Ferdinand Samare was appointed as Minister for Commerce and Tourism, further obscuring the division between Nimamar and 'official politics'. Lihirians were growing eager for their own government, and by 1988, local councillors and Nimamar members discussed the idea of breaking away from the Namatanai Local Government Council. Lihirians considered the possibility of an island local government which would include both Tanga and Anir, but not the Namatanai mainland — a notion entirely unfathomable in the current context of hostility towards any economic or political alignment with other New Irelanders (Bainton 2009). In 1988, the New Ireland Provincial Government passed a *Community Government Act*, largely to appease Lihirians after the initial feasibility of the mine had been established. The Nimamar Community Government was established under this legislation, representing the first real transition towards a more autonomous form of local governance. This might confirm Worsley's (1968) Marxist theories of political trajectory, were it not for the fact that various supernatural and millenarian beliefs have remained a stable feature of the Nimamar movement, even influencing the ways in which some younger people regard contemporary politics and development issues. If it appears as though these beliefs have been disposed of, in so far as they no longer feature in daily discourse, it is probably more likely that they have been (temporarily) displaced by new concerns that arise through the unpleasant experiences of development as it is delivered through mining operations.

As mining negotiations continued into the 1990s, Lihirians were determined that they would retain a greater proportion of the potential economic benefits. The New Ireland Provincial Government passed a bill in 1994 to transform the Nimamar Community Government into the Nimamar Development Authority (NDA), with an increase in the number of wards from 12 to 15, and the introduction of Ward Development Committees and Village Planning Committees. The NDA was supposed to function as a type of interim local-level government prior to passage of national amendments to the *Organic Law on Provincial Governments and Local-level Governments* in 1995, which happened shortly after the signing of the IBP agreement. Under that agreement, the NDA was granted control of funds to be spent on community development projects in Lihir (Filer 2004: 3). It was not until 1997 that Lihirian leaders held the first elections to officially reconstitute the NDA as the Nimamar Rural Local-Level Government (NRLLG). The delay may well have been the result of internal divisions between supporters of the government and those loyal to older versions of Nimamar. It may also have been due to the role of the NDA as an administrative unit, rather than a political body, during the project construction phase. In any event, the political inactivity of the NDA opened the door for the Lihir Mining Area Landowners Association (LMALA) to emerge as the major political force on the island, largely as a result of the all-encompassing political struggles between the mining company and newly relocated landowners.

The 15 Ward Members in the NRLLG elected Clement Dardar as their Chairman, although he insisted on being called the President. Only two members of the old Nimamar group were re-elected as Ward Members. While the original NDA officially remained as the administrative arm of the NRLLG, Lihirians continued to refer to their local authority as the NDA, which may have been a reflection of public confusion about the role of the NRLLG, which was now in control of substantial amounts of money derived from the mining project that were intended for 'community projects'. Problems surrounding the relationship between the NDA and the NRLLG were not resolved until 2001, when the Nimamar Special Purposes Authority was established to replace the NDA as the body that would administer NRLLG-funded projects. The original Nimamar members were steadily replaced as Lihirians became more involved with mining activities. However, strong anti-State sentiments and millenarian aspirations have remained influential within the NRLLG and throughout the wider Lihirian community.

New Order

Although Lihirian forms of political change have their roots in the early New Hanover movement, in many ways the post-war Paliau movement in the Admiralty Islands (now Manus Province) provides a more useful analogy for the contemporary political process in Lihir, highlighting the way in which cultic movements are generated by some form 'extreme' experience. Even though Lihirians already had such experiences in the colonial period, the mining project has thrown this experience into sharp relief and given it a new life.

Both Mead (1956) and Schwartz (1962) have closely documented the social, cultural, political, economic and religious transformations that took place in Manus in the years following the Second World War. The overwhelming experience of the war involved the introduction of local people to new technology delivered by the American forces on a huge scale, and the observation of comparatively equal relations between White and Negro soldiers, which suggested that existing colonial lines between *masta* and *boi* could be dissolved. This left Manus people with the feeling that their lives had been irrevocably altered. Paliau, a former policeman, came back to Manus after the war with plans for a 'New Way' (*Niupela Pasin*), that envisaged a break with traditional social organisation and religion in favour of new village formations, organised (and often highly ritualised) communal work and savings, and the establishment of schools, councils and village courts. The movement emphasised social unity and attempted to bring together previously conflicting groups throughout the Manus region, yet unlike the Lihirian movements, there was a positive attitude towards the government and a strange antipathy towards the Christian missions.

In 1947, and again in 1952, many supporters of Paliau were caught in the grip of cargo cult movements that emerged in competition with the Paliau movement. Cult followers anticipated a 'Second Coming' of Christ, destroyed property (or anything associated with the 'old ways' that might be seen to 'block the path' for delivery of the new cargo), engaged in secondary burial rituals, and experienced visions, supposed miracles, and shaking (the so-called 'Noise'). Paliau tried to resist these cult manifestations, but as his own expectations and prophesies failed to materialise, he was unable to capture the growing numbers of cult followers, and this gradually led to the decline of the Paliau movement. However, in some ways Paliau and his supporters did achieve a desired level of change. The Baluan Local Government Council was established in 1950, Paliau became its President, and he eventually went on to become a member of the House of Assembly and gain the respect of the colonial administration.

While there are obvious parallels in Lihirian reactions to colonialism and local political development, the more telling comparison is found in the ways in which Lihirians have responded to mining. As we shall see in Chapters 6 and 7, the Destiny Plan, which involved the reconstruction of cosmology and the reproduction of mimesis and insular political views, the efflorescence of *kastom* (that is simultaneously about individual pursuits and greater social unity), and the increasingly legalistic discourse surrounding it — they have all emerged in response to the radical and abrupt experience of mining.

Although Lihirians have actively directed the course of events in their history, their actions have not always produced the desired outcomes. They have responded to marginality in multiple ways, all of which express dissatisfaction with social inequality. As their aspirations for moral equality and material wealth were consistently denied, and their ritual means for achieving these goals proved inefficacious, the result was an even greater antipathy towards the colonial administration and later the national government.

To be sure, we can recognise an internal cohesion within TKA and Nimamar beliefs and interpretations (see Horton 1970). But from a Western reductionist perspective — or what Weber would regard as the 'specific and peculiar rationalism' of the West — what was rational *to* these Lihirians was not necessarily rational *for* them — that is to say, the stated means were irreconcilable with the desired ends. Ever since F.E. Williams (1979) characterised cargo cults as a form of 'madness', anthropologists have tended to side-step their 'non-rational' aspects through functional analysis. As a result, there is a common tendency to emphasise the socially positive cultural consistencies or internal rationality of the associated beliefs, effectively censoring the judgment that people perceive themselves as effecting particular ends through activities that ultimately fail.[16] This is an important point, because in the following chapters we shall see how a new group of elite Lihirian leaders employed a form of bourgeois rationality to critique and reject Lihirian cultural interpretations — the 'mythologies' of wealth — only to reinvent these 'myth-dreams' within a more complex scenario.

Lihirian responses to inequality are regularly vocalised and expressed through anger directed at the national government and the mining company. Lihirians have not internalised their concerns and they do not exist in an abject state. However, the ways in which they set out to achieve their ends are varied and diffuse. Lihirians have regularly employed numerous political strategies (which are often contradictory) when dealing with the government or the company. Understanding these responses to inequality, the demands and expectations which people place on the company, and the more general way in which they 'menace the mining industry' (Filer 1998), means acknowledging the unholy

---

16   There are some exceptions. Andrew Lattas (2007: 158) argues that, instead of seeing 'madness' or irrationality as the result of inconsistent Western policies, or contradictions in Western ideology or values, we should consider this appearance as the result of cultural myths and ontological schemes in which transgression is figured as a creative act.

trinity between resource dependency, the apocalyptic elements of Lihirian Christianity, and the secular decline in the legitimacy of the post-colonial regime.

The mining project with its related infrastructure, services, opportunities for wage labour, and payment of royalty and compensation monies — what Lihirians have come to recognise as development — has been generically interpreted as the fulfilment of prophesies, but for many this is not the new millennium. For those who retrospectively interpreted Arau's prophesies as the forecasting of a Lihirian 'city', many of the more pernicious aspects of their urban aspirations have come to pass. But for those who hoped for a just and equitable future, a virtuous society of equals in which everyone would be rich, recent experiences may simply represent their flight into the maelstrom of modernity.

4. Lihir Custom as an Ethnographic Subject

The cultures of New Ireland have long held a certain level of anthropological attention, which is largely due to the Western fascination with the elaborate mortuary rituals and the production, form, use and iconography of *malanggan* mortuary carvings found in the northern part of the province.[1] The Lihir Islands are situated between the ethnographically better known islands of Tabar and Tanga, and are easily visible from central mainland New Ireland, where there has also been extensive ethnographic documentation, yet somehow Lihir remained comparatively blank on the ethnographic map. However, since mining activities began, Lihir has been more 'anthropologised' than any other part of New Ireland, with attention mainly directed to understanding the social changes associated with such activities.[2] In this chapter, I provide an outline of Lihirian *kastom* as it appeared in the period before mining. Any description of Lihirian culture cannot be detached from history, and as we have already seen, there was a pronounced articulation between Lihirian culture and external forces during the previous century. However, the cultural shifts experienced during that period were relatively gradual, and were generally not accompanied by the feeling of dislocation, whereas the transitions that have taken place since mining activities started have produced a profound sense of cultural rupture. Nevertheless, as we shall see in Chapter 7, Lihirians have responded to mining in a rather cult-like fashion that has generated an efflorescence of custom, using newly acquired resources to ensure that *kastom* remains relevant and strong in an entirely altered context.

These cultural adjustments have also been accompanied by broader social changes. Some Lihirians have recalibrated relationships and 'cut' their networks with people in neighbouring areas — and to some extent across Lihir itself — in order to contain the distribution of mine-related benefits. This has given rise to

---

1   *Malanggan* carvings include figures, masks and large woven disks that are produced and used as part of the mortuary ritual cycle (see Lincoln 1987; Parkinson 1999; Küchler 2002; Gunn and Peltier 2006; Billings 2007).

2   Since exploration started, there is a long list of social scientists who have conducted research projects on Lihir, all of which involved fieldwork that concentrated on some aspect of Lihirian culture. Consultants have been engaged to undertake social impact studies and genealogical investigations (cited in Chapter 2), to provide advice on the management of community affairs, the design of relocation exercises and community development projects, and more recently to develop a cultural heritage management program (Bainton et al. forthcoming b). Most of these consultants have published some of their findings. There have also been at least nine independent research projects that involved fieldwork in Lihir, resulting in one Pre-Doctoral Diploma, four Masters and three Ph.D. theses (Skalnik 1989; Awart 1996; Benyon 1996; Lagisa 1997; Kowal 1999; Hemer 2001; Bainton 2006; Burley 2010; Haro 2010). In addition, LGL has directly employed two expatriate anthropologists within its Community Liaison Department and established a Cultural Information Office that employs a Lihirian graduate anthropologist.

a distinct sense of ethnic difference which is in some respects a crude extension of the distinction between 'landowners' and 'non-landowners' (Bainton 2009; see also Nash and Ogan 1990). To some degree, the efflorescence of *kastom* is also contingent on these changes. The desire for greater unity in the face of increasing atomisation, and pressure to redistribute mining wealth across the wider community, has simultaneously transformed existing cultural practices, increasing the competitive elements of *kastom*, while the ideology of *kastom* places a greater rhetorical emphasis upon unification.

In light of these recent transformations, it is worth noting the extent to which different groups throughout New Ireland were historically inter-dependent. Indeed, it is more appropriate to understand Lihir as part of an 'areal culture' than as an isolated 'cultural area' (Schwartz 1963). Groups from Lihir, Tanga, Namatanai, Lelet, Lesu and Tabar are integrated through matrilineal descent and kinship relations, leadership structures, the men's house institution, mortuary rituals, and traditional forms of trade and exchange. Regardless of scale, the boundaries of any areal culture will always be fuzzy (Knauft 1999: 6). Despite political atomism, and a corresponding particularism in all institutional spheres, these New Ireland groups are socially and culturally integrated to the extent that they constitute a single areal culture, and each must be considered as a 'part-culture' which is neither self-sufficient nor fully comprehensible by itself.

Birds of Afar?

At the most superficial level, Lihir is a matrilineal society divided into two exogamous moieties which bear the vernacular names *tumbawin-lam* (big people cluster) and *tumbawin-malkok* (small people cluster). *Tumbawin* is a generic term that literally refers to a bunch (*tum*) of bananas (*win*), and is often used to refer to all groups of people — be they moieties, clans, sub-clans or lineages (see Wagner 1986: 84–6). The metaphorical reference to a bunch of bananas conveys the idea of a group with a common ancestress that is already in the process of dividing into separate branches, if it has not done so already. At some point in the past century Lihirians adopted the New Ireland Tok Pisin names Bik Pisin (big bird) and Smol Pisin (small bird), identified with the White-bellied Sea-eagle (*Haliaeetus leucogaster*) and the Brahminy Kite (*Haliastur indus*). It is likely that these terms were introduced through marriage connections to mainland New Ireland. In Lihir these terms may also be used to simply mean 'signs', because unlike their mainland counterparts, there is no strong identification with these 'totemic' markers. There are numerous theories and myths that account for the origin of the moieties.[3] However, the totemic designators cannot be understood in the Durkheimian sense of the term. Lihirians do not imagine a sacred or ritual relationship between a social group and its 'totem' as an emblem of membership and the focus of social solidarity. This may be an instance of what Kroeber (1938: 305) once described as a situation in which totems are merely epiphenomenal.

The practical significance of the moieties was mainly limited to marriage preferences, although genealogical records suggest that the moiety system was compromised at some point in the past, so the principle of dual organisation no longer served to regulate marriage.[4] Although most Lihirians insist that there were once strict prohibitions on intra-moiety marriage, the taboo now applies to marriage within the clan.[5] Eves (1998: 123) notes that, among the Lelet people (on the New Ireland mainland), moieties might be the largest and most fundamental kinship category, but they are not corporate groups; they are the means by which marriage is ordered and structured, not the means whereby people gain access to land, resources, valuables or property (this is the role of the clan). Wagner (1986: 51) argues that Barok moieties (also on the New Ireland mainland and closer to Lihir) cannot adjudicate or settle matters, and 'are less a means of organisation than one purely of elicitation', as exchanges and alliances between the two groups elicit important social values. Both of these arguments can be applied to Lihirian moieties; for the most part, they do not play an overly practical role in Lihirian social organisation.

Social Units

In Lihir, the pre-capitalist (and pre-mining) concept of the clan is of a social unit that fuses three core symbolic elements: *a le* (shell money), *rihri* (men's house), and *tandal*, otherwise known throughout New Ireland by the Tok Pisin term *masalai* (spiritual beings that inhabit the cultural landscape).[6] While

---

3   Powdermaker (1971: 34) notes a similar situation among the Lesu of the New Ireland mainland, but mentions that the origin of the moieties and clans can be found in stories about an ancestral heroine.

4   This complex system, often referred to as the 'classificatory system', is readily accounted for by dual organisation. However, as Levi-Strauss (1969: 72) points out, it is the principle of reciprocity, not dual organisation itself that constitutes the origin of the classificatory system. Anecdotal evidence suggests that Lihirian moieties were at one stage connected by reciprocal prestations and counter-prestations through the feasts and exchanges associated with mortuary rituals.

5   Unlike people from Lesu or Lelet, Lihirians do not readily recall times when breaching the rules of moiety exogamy were met with violent death at the hands of kinsmen (Bell 1938: 318; Powdermaker 1971: 41; Eves 1998: 133). As early as the 1930s, Groves (1934: 239) notes that it was common in Tabar to see marriages between individuals belonging to the same moiety, and that these were not regarded with particular 'disfavour' if there was sufficient residential distance or no traceable degree of consanguineous connection between the spouses.

6   A quick glance through the Melanesian literature soon reveals that the Tok Pisin term *masalai* has a wide variety of local connotations, covering various permutations of what can generally be regarded as 'bush spirits', yet in each place people have very specific ideas about what these are (see Lawrence and Meggitt 1965).

the nature of Lihirian clans is culturally determined, in so far as they exist they are units of social structure. Big-men were necessarily implicated in this symbolic trinity as the 'owners' and leaders of the men's houses, the mediators and human counterparts of the *tandal*, the custodians of each clan's store of *a le*, the arbiters of protocol and the organisers and chief orators at feasts held within the men's houses. It was impossible to conceive of big-men outside of the cultural conception of the clan and vice versa. Through the autonomous actions of big-men, these elements gained symbolic and practical significance, drawing people together in ways that created and maintained group unity and identity. This image is structurally similar to Wagner's (1986: 84) depiction of Usen Barok clans, but in both cases this no longer adequately captures the dynamics of contemporary leadership or group organisation.

The structure of Lihirian clans has been transformed since the inception of mining activities, particularly as the distribution of mine-related wealth (ideally) occurs along clan lines, ultimately defining clan membership according to the eligibility to receive such wealth. The boundaries of Lihirian clans have always been extremely permeable and 'thick', not clear and incontrovertible. What Ernst (1999) calls the 'entification' of clans as a result of mining has meant that relationships with non-Lihirians have become more circumscribed, and incorporation of outsiders has become the source of some tension.

Although Lihirians often speak of a 'clan system', it has proven exceptionally difficult to reach consensus on how the various clans, sub-clans and lineages are systematically linked. However, insistence from all Lihirians that the clans can 'fit together', despite inconsistencies between all the models on offer, indicates the extent to which Lihirians have come to believe in the ideal of 'organised' group membership and unity. Regardless of the difficulty of working out how or whether all of the clans connect in a cohesive way, the important thing is that Lihirian clans are not recent fabrications made out of economic and political necessity for the distribution of mining benefits (Filer 1997a: 177). On the other hand, there is a sense in which Lihirian clans have become bureaucratic units organised around the administration of mine-derived wealth, with some (re) forming themselves as businesses and corporations built around contracts with the company.

After presenting four versions of the 'clan system' in their social and economic impact study for the mine, Filer and Jackson (1989: 55) concluded that 'local people's models of the system are as diverse as the interests which they express', and that the 'system' as a whole is 'best understood as an ideal world which must be constantly adjusted to a set of personal relationships whose instability it is intended to conceal'. Given the economic and political changes that have since taken place, this observation is doubly pertinent.

In the late 1960s, Yngvar Ramstad identified six major clans in Kunaie village. He acknowledged that some clans may be known by different names in other parts of Lihir, and that the major clans were best understood through the local idiom of the Canarium nut, the kernel of which contains several parts, illustrating how numerous clans can be included or 'covered' under the one encompassing clan name. When viewed from one or two villages, it is easier to make sense of the relationship and ordering of the clans: the number is usually limited, and there is more likely to be consistency between various versions of the 'structure'. On the small island of Mahur, Susan Hemer (2001: 16) found that isolation, a smaller population and fewer clans made it easier to comprehend how groups were ordered. However, my own investigations have only multiplied the number of models on offer and established that there is no single authorised version. Given that my research was not restricted to a single area, instead reflecting the high level of mobility in Lihir, it was only when I considered Kinami village as an isolated area that I could draw limited conclusions about how the clans 'fit' together. Difficulties arise when Lihir is considered as a whole — as a group of five islands, some 35 villages and over 400 hamlets.[7]

The 'official' version used by the Lihir landowners association (LMALA) contains six major clans divided between the two moieties, which contain approximately 70 sub-clans. Many of these are further divided into lineages, although some lineages can be considered sub-clans in their own right. However, the concept of six major clans is partly a political fiction that lumps groups together for bureaucratic organisation and collective representation. As a result, this model has been heavily contested. In most instances, it appears that Lihirians only maintain a basic comprehension of the clans and sub-clans to which they are immediately related. This enables people to work out their own position in relation to other people and clans, and more importantly to know what resources they can access.

Perhaps it is better to understand the 'system' as 'organic', in that it appears to have adapted to changes in population size and settlement patterns over time. It does not possess an easily recognisable two-dimensional structure in which clans and sub-clans are evenly distributed between the two moieties, with clear lines of relationship between individual groups. Lihirians can usually only recall ancestors two generations up, meaning that genealogical knowledge is quite shallow (see Foster 1995a: 73). This has some implications for reconstructing

---

7   The number of hamlets is taken from maps produced and used by Lihir Gold Limited in 2004. In recent years, Luke Kabariu from Masahet Island, who works as the Cultural Information Superintendent in LGL's Community Liaison Department, has attempted to synthesise the various models into a definitive account. Although there is some level of community consensus about the overall structure, not everyone accepts all of the relationship lines. Given the highly politicised nature of clan relationships within the context of mining, I have not attempted to include the variety of structures or relationships, or present a synthetic example that may later be construed as reality or taken as a single authoritative version.

the ways in which the different clans are related to each other. Anomalies may be the result of an earlier breakdown due to a small population, or because the system of dual organisation (moieties) is a foreign structure imported and imposed on an already asymmetrical arrangement.

A number of Lihirian clans are represented in language groups from Barok, Patpatar, Susurunga, Tabar and Tanga, where moiety division is a common structuring principle for marriage and exchange (George and Lewis 1985: 32; Clay 1986: 55). Moieties may well have been 'introduced' or 'adopted' from alliances with these groups, through which Lihirians also gained access to land and resources. There may be cultural myths which attempt to explain the presence of moieties, but as Levi-Strauss observes, we are likely to remain ignorant of their origin, and the 'problem of whether clan organisation resulted from a sub-division of moieties, or whether moieties were formed by an aggregation of clans' continues to have little importance (Levi-Strauss 1969: 74). What we find throughout most of the southeastern part of New Ireland are 'differences of degree, not differences of kind; of generality, and not of type' (ibid.: 75). Lihirian patterns of social organisation reflect those found throughout the region, and although people subscribe to the structuring principles of dual organisation as a stated ideology, there are relatively few domains where it affects their practice.

The variety of local opinions does not imply the fragmentation of a once shared understanding, nor that the system is now somehow broken. Rather, it reflects a common Melanesian pattern of fluid social organisation and group identification. Therefore, it is more informative to consider the types of relationships between groups that are expressed in clan origin stories. These are often a mixture of metaphors and history that describe how certain clans were created by others or came into a partner-like relationship, and recount the migration of people and the mothering of new lines as people and groups broke away from each other for various reasons. Origin stories reveal at least four types of relationship: identity, inclusion, partnership and separation. Some depictions reflect how the same clan can be known by different names throughout the islands, while others emphasise a relationship of identity or show how larger clans encompass, include or 'cover' a set of smaller clans. Many of these stories also reveal a strong attachment to place and serve to illuminate the Lihirian cultural landscape. Place names act as mnemonics for historical movement and action, emphasising the relational qualities of the landscape, and prompting our comprehension of landscape as cultural process and the more phenomenological understanding of place as *event* (Weiner 1991; Casey 1996).

In some instances, these origin stories conform to the actual relationship between clans and sub-clans. In other instances they contradict one another. It is then difficult to decipher whether particular clans have a certain relationship with one another because of this 'pre-existing relationship', or whether certain events, or new alliances or splits, have generated a new set of relationships, around which people reconstruct their image of the system. Disparities arise because different people consider the relationships between certain clans in different ways. Many of the stories are disappearing, particularly as the tradition of transmitting knowledge to younger generations is in decline, which will only render future attempts to reconstruct the system increasingly futile.

Collective Rights

The principal benefit of clan membership is the right to access clan land and resources in order to subsist. There are several main categories of customary land rights: clan land, lineage land, individually owned land, and land provided for newcomers or adopted clan members. Lihirian clans consist of a large number of lineages which may be dispersed across the islands. This means that collective rights to cultivation or use are generally associated with the lineage, and they are distributed, in practice, by the senior males in each one. By contrast, the ownership or transaction of land generally occurs at the clan or sub-clan level, and ideally any transactions or decisions must have the consent of all significant clan leaders and be publicly agreed upon. Clan land generally includes bush and gardening land, beaches or fringing reefs, and sacred sites that are usually prominent physical features such as rocks or water sources. Lineage land tends to be more specific, often including hamlet areas, men's house sites, and gardening land.

Land rights have always been intricately connected to feasting and exchange, which is the most important avenue for securing rights within the clan. Land which is exchanged or 'purchased' in the context of customary feasting and exchange is generally held by individuals in perpetuity and inherited by specific matrilineal heirs. In the past, inter-clan exchange debts (and in some cases compensation debts) could be settled with land transactions. With the commoditisation of land, these practices have become increasingly rare because clans are increasingly reluctant to part with potentially valuable resources. Due to a long history of outside influences, Lihirians have come to view traditional land rights as neat parcels that can be handed out to clan members, reflecting the hardening of previously flexible group boundaries and the more precise delineation of group territories.

Primary rights include access to clan resources and land. Adult clan members generally have equal access to, and proprietary claims over, garden land and residential hamlet land. Individuals have rights of access to clan land for purposes such as gardening, hunting, and gathering food, timber and other materials. Although individuals can claim damages for destruction of land

and resources that they own, this process will generally be arbitrated by clan leaders. Secondary rights are those of usufruct, where individuals can use land and have certain rights over trees or gardens on that land. In-marrying people will generally be granted usufruct rights over the lineage or clan land of their spouses, and other immigrants can be granted usufruct rights on their acceptance as residents in a new area. Secondary rights can be converted to primary rights if the person in question has been accepted as a member of the clan and has significantly contributed to feasts with gifts of pigs, shell money and cash. Ideally, these rights must be conferred by all members of the clan who decide on the incorporation of the newcomer.

In the distant past, there was a practice known as *erkuet* by which a clan leader might instruct his fellow clansmen to strangle his own wife after his death so that she might be buried with him as a mark of respect for her own involvement in the feasts and exchanges that produced his high status (for similar practices, see Groves 1935; Chowning 1974; Goodale 1985). The husband's clan would then provide her clan with a parcel of land as compensation for her death, which was then considered to be inalienable. Some men have suggested that widows happily submitted out of love for their husband. Women's accounts indicate there was some coercion by the husband's brothers, who feared that the wife might remarry and thus transfer any status, knowledge or wealth she had accumulated through her first husband to a rival clan. Regardless of the motivation, this form of transfer is no longer an option, and the certainty of title which it is believed to have had can never be tested, which is somewhat different from transactions that arise through customary feasting (Burton 1993, cited in Filer 2006).

While there are instances where fathers can give land to their children, as happens when children hold the customary *ikineitz* feasts for their fathers, land is normally inherited through the female line. Since mining activities began, there has been a growing tendency to 'bend the rules' of inheritance as fathers increasingly concentrate on their own children's needs at the expense of their sisters' children, or as people attempt to isolate parcels of land from the customary system of inheritance. While mining has produced the most profound impacts on Lihirian land tenure and ideology, the emergence and acceptance of individualistic (or at least restricted) 'ownership' owes its origins to earlier cash cropping ventures that encouraged individual holdings, and also to missionary emphasis on the nuclear family.

Not surprisingly, Lihir reflects a typical Melanesian pattern of chronic fission and fusion between individuals and groups of various sizes. The colonial administration conveniently construed this reality as a rigid expression of customary land rights, by abstracting things like social groups and land boundaries from the existing social landscape. The irony of the situation is that

Lihirians have since adopted a similar conception of their 'clan system' and the rules of land tenure. Despite administrative folly, or local assumptions that owners and rules are easily identified, it is the inherent flexibility of customary land tenure that allows Lihirians to adapt the forces of development for their own purposes, which also makes formal delineation all the more difficult. Tensions have emerged due to a reluctance to acknowledge the extent to which contemporary land tenure practices differ from those of a supposedly immemorial tradition (see Ward and Kingdon 1995: 36; Weiner and Glaskin 2007). Local attempts to maintain an image of an idealised traditional system continually confront the fact that land has become a commodity.

Organising 'Principles'

Clans (*tumbawin*) and sub-clans (*tsiretumbawin*) are comprised of lineages (*bior*) each of which can theoretically trace descent from a common female ancestor or point of origin. When people are born, they are immediately accepted as members of their mother's clan and the men's house which is looked after by her brothers (of which she is also member). This aspect of one's identity is axiomatic and immutable.[8] People differentiate between members of a single clan through lineage affiliation. In reality, a lineage is that group of people whose lives revolve around a particular men's house, who work and garden together, and come together to host various feasts. Genealogical knowledge of the lineage tends to be lateral rather than vertical: people who were not renowned as leaders or for some other prominent reason are sometimes dropped from the collective memory.

Lineage identity might be fixed at birth, but in practice membership is defined performatively rather than genealogically. This is partly attributable to the fact of co-residence, as people within a single hamlet pool their labour in gardening and custom work. The flexibility of residence attests to the 'openness' of Lihirian ways of calculating relatedness. Thus people living away from their own land, sometimes because of marriage arrangements, may find that, for all practical purposes, they have become incorporated into another lineage. This similarly applies to people (both Lihirians and non-Lihirians) who have been adopted by other families: they are expected to contribute to the well-being and maintenance of the adoptive group.

---

8   Lihirians also refer to lineages by the term *dal wana pour* (matrilineal blood line). In addition, people regularly use the Tok Pisin term *famili* (family). This can prove confusing, given that *famili* is also used to describe the nuclear family unit, including the husband, who is certainly not a member of his wife's matrilineage. This partly reflects education and social changes that place greater emphasis on the nuclear family. Nevertheless, identity — and more importantly, access to resources — is still primarily constructed in terms of matrilineal reckoning.

People who arrive in Lihirian villages, hamlets or men's houses, regardless of whether they are Lihirians from another village, or non-Lihirians from another area, are generally called *wasier*, although there have been recent changes in the ways that people regard outsiders. In the vernacular, the term means guest or visitor. *Wasier* differ from members of the household or lineage because they are seen as recipients of the work and hospitality of the household or the lineage associated with a particular men's house. Implicit in the use of this term is a sense of obligation to *pinari wasier* (to provide food, hospitality and gifts for guests) — a vital function of households and men's houses. *Wasier* are an accepted category of relations across the spectrum of villages and islands, and traditionally *wasier* from neighbouring areas knew that they would be assured of hospitality, and if they wanted, they could stay on as incorporated guests.

The fluidity of relatedness is illustrated by the ease with which people can reside in other hamlets. Often people will be closer to those with whom they are in daily contact, meaning that, even though there might be other kinship categories that stipulate a close relationship, if these people are not present then closer bonds will be established with those who have a more direct bearing on people's daily life. However, it is also common for the 'relationship' to overrule residence; distant relatives often appeal to the relationship category in order to establish claims for assistance. This means that, despite the flexibility of incorporation, there are certain kinship categories that carry greater purchase on the demand or obligation to render gifts.

Individual Relations

The structures of Lihirian kinship generally resemble those of neighbouring communities. While there are minor variations, the most prominent relationships common to this region of New Ireland involve a strong connection with maternal uncles, cross-cousin relationships marked by forbearance and conviviality, avoidance between cross-sex uterine siblings, and relationships between in-laws that are constrained but highly respectful.

The strongest relationships in Lihir are those between a man and his sister's children. Mother's brothers are referred to as *motung* by their nieces (*liling wehien*) and nephews (*nunglik*). Mother's brothers are supposedly the most important men in a child's life, and the child can expect nurturance, guidance, assistance and discipline from them. Ideally, they will pay a man's bridewealth and provide him with access to clan land and resources. Traditionally, *motung* would hold and control any shell money or cash gained by their nieces and nephews, which is a good reason why the mother's brothers were regarded as their 'big-men' (*tohie*). For males, this was a relationship of co-dependency: younger men relied upon their uncles for their own growth, knowledge, development and support (for both cash and shell valuables), while older males relied upon their nephews for labour, political support and lineage succession. Younger age mates existed in symmetrical relationships of relative equality until they were old enough to begin displaying personal qualities that set them apart from others. Senior men exerted control over younger male labour power, time and economic resources.

Women were also subject to direction by their male kin and their husbands. Despite the payment of bridewealth (*rapar*), there is a tension between a woman's brothers and her husband. Brothers often attempt to keep control of any wealth gained by a sister, fearing that her husband may convince her to part with it and hence place it at the disposal of another clan. This gendered hierarchy is played out in numerous situations where men assert their influence and authority over women in daily affairs. In contemporary contexts, this dominance is often maintained through appeals to tradition and *kastom*.

Lihirian cross-cousin relationships exemplify classic textbook joking relationships. These are most obviously contrasted with the strict avoidance between uterine brother and sister (see Bell 1935a; Clay 1977: 47; Wagner 1986: 70) and the sense of competition and latent tension between uterine brothers. The level of competition between brothers is not matched by similar competition between sisters because, unlike brothers, sisters are not vying for leadership positions within the lineage. Relational strains emerge from political pursuits: essentially brothers utilise the same resources in their competition for influence and esteem. Given that brothers are united under the influence of the same man (their senior *motung*), jealousies inevitably arise as one brother gains more favour than another. Jealousy between brothers is said to have complicated the relationship between kinship and residence, which has continued to discourage them from settling in the same hamlet. Previously, it was typical to find within a hamlet one senior male member of the lineage, who usually 'owned' the men's house, while the remaining male residents were related as either sons (*zik*) or sister's sons (*nunglik*), or even as sister's husbands or sister's daughters husbands.[9] Previously, high levels of village endogamy meant that brothers were not split by marriage, and younger nephews were in constant contact with their uncles.

The close bond between cross-cousins and the idealisation of this relationship must be seen against the background of uterine sibling relationships. Cross-cousins are typically separated by belonging to different moieties or at least different clans. They exist under the influence of men who are related as affines and avoid one another. The relational distance between cross-cousins, and the

---

9   This is only a general outline as there are numerous qualifications. In some cases, where there are no brothers or sister's sons to succeed to ownership of the hamlet, the son of the current owner may then become the senior male of the hamlet, even though he does not belong to that clan, or technically 'own' the men's house. In such cases, he looks after the men's house until younger males in the clan can take over.

existence of different authority figures, means that they are not likely to be divided by jealousy or competition. These links are extremely strong, and it is particularly difficult to resist demands and requests from cross-cousins — whether they are for betelnut, money, beer, or in one sordid case I was told of, assistance with a pack rape. In times of warfare, one could supposedly seek refuge behind a cross-cousin and be assured of safety, because even enemies recognised the importance of this relationship. Cross-sex cross-cousins are similarly close, and this relationship is equally characterised by joking and conviviality and stands in marked contrast to opposite-sex uterine sibling relationships. Between opposite-sex cross-cousins there is a taboo on the use of personal names and kinship names are used instead. In the past, cross-cousins who were not first cousins were seen as ideal marriage partners, which was consistent with the practice of moiety exogamy.

There is a close bond between fathers and their children that was said to be expressed through their relationship with their father's *tandal*. It is believed that when children, especially sons, ask their father's *tandal* for assistance it will oblige. In addition to feasting and exchange relations, these relationships unite people across moieties and clans. In the past, relationships between in-laws (*poas* in the vernacular; *tambu* in Tok Pisin) were more circumscribed, involving the strict avoidance of names and not entering the other's house. Females had to avoid being physically above, or walking directly past, the heads of their male in-laws if the latter were lying down. This relationship was not symmetrical: it is the spouse who had to show respect towards their in-laws. A man generally avoided or felt shame towards those people who shared a part of the bridewealth paid on his behalf. As a result, people might find reasons not to accept part of the bridewealth in order to maintain a pre-existing relationship. Likewise, some would take the opportunity presented by marriage to assume a relationship of formal avoidance in order to alleviate difficult or stressed relationships. As Clay (1977: 43) remarks of similar Mandak practices, 'people often have a choice among two or more alternatives in determining a particular social category, and individuals often seem to emphasise certain relationships for political purposes'.

Regardless of firmly stated ideas, in practice there is variation. It is becoming evident that many of the former rules of avoidance are no longer adhered to by younger generations or enforced by their elders. People are active agents in relationships: kinship sets the boundaries or the field of action, often establishing what is impossible, possible or probable, but it cannot wholly determine the decisions that individuals will make or the character of a given relationship.

Dead Men and Persistent Values

It may appear like a tired truism to state that Lihirian kinship and traditional politics (by which I mean male leadership roles and mortuary feasting), both as ideologies and contexts for meaningful action, are coterminous with one another. Fortes and Evans-Pritchard (1940) made this point in relation to African kinship and political systems, and Allen (1984) has reiterated the same point for Melanesia. In consideration of Lihirian male leadership and sociality, it is worth noting the extent to which both kinship and politics were intimately bound — a point which has assumed greater salience as men now seek to divorce their kinship obligations from their personal, political and economic aspirations.

Filer and Jackson (1989: 185) concluded that Lihirian big-men, for the most part, were essentially 'dead men'. This might seem overly pessimistic if not for the fact that the sort of leaders who could actually qualify as big-men ceased to exist when the traditional economy which bolstered their position was subsumed by capitalism. The ritual and esoteric knowledge they alone possessed no longer provided appropriate responses to their changing environment. While Lihirians often claim that they have contemporary big-men, and readily point out men believed to possess the requisite qualities, the proliferation of leadership types has meant that there is little unanimity on the question of what is a big-man (Bainton 2008).

Male and Female Status

Lihirian leaders are referred to in the vernacular by the term *tohie*, which finds translation in Tok Pisin as *bikman* (big-man). Although *tohie* is used as a title of status, it is also used generically to mean leader. It is not the sole reserve of males, as it can be applied to senior women who have proven themselves in customary feasting and exchange, and where they are the sole surviving senior member of the lineage and have thus become the 'owner' of the lineage's men's house. These women are described as *wehientohie* (big woman), and in the past some wore shell armbands (*kual*) to signify their status.

Although women can achieve status and renown within their clan, this does not mean that they enjoy the same level of influence accorded to senior male leaders. While they might be recognised as the 'owner' of the lineage's men's house by default, this similarly does not equate to additional rights within the men's house. Given that most discussions take place within the men's house, women are generally excluded from taking an active role in clan and lineage affairs, which contrasts with the situation described by Nash for the Nagovisi of Bougainville (Nash 1987).

Unlike male status, which is cumulative, female status is two-staged. There are no male initiation ceremonies or graded societies, although there were numerous secret societies, such as *Pindik*, *Triu* and the *Tanori* fishing society. Men gradually achieve higher rank through political and economic endeavour, whereas women are generally seen as mature or immature, based on their ability to carry out tasks such as cooking, gardening, rearing children and raising pigs. Although men and women contribute in different ways to gardening work and raising pigs, in the past this work was highly valued. Previously, pubescent women were initiated into womanhood through the *tolup* or *ba'at* ritual, in which they were kept in seclusion for some months before being married, but since missionisation this practice has been abandoned.[10] While mining has definitely provided new opportunities for female leadership and autonomy (see Macintyre 2003, 2006), generally women remain subject to male leadership.

Abilities and Possessions

Traditionally, big-men were rich in land, pigs and shell money. They were outstanding warriors, courageous, excellent orators, and the guardians of ritual and esoteric knowledge. These positions were not hereditary, conforming to other images of big-men throughout the region, where the office of leadership is a personal achievement (see especially Oliver 1955; Powdermaker 1971; Clay 1986; Wagner 1986; Foster 1995a; Eves 1998), although first born sons are accorded an important status (*ziktohie*) and being the son or nephew of a successful big-man may provide an ambitious young man with extra resources and a platform upon which to launch his own career. Nevertheless, theoretically every man is eligible for bigmanship.

In the past a clan's leading big-man was the custodian of its store of shell money, and he controlled the distribution and use of all pigs and garden produce. It is hard to assess the leader's reach and control within the clan, even though elderly men argue that the authority of the clan head was absolute. This is a difficult image to reconcile with contemporary leadership practices and the extensive splitting of all Lihirian clans. Needless to say, this style of leadership has been long contested by younger men determined to assert their own autonomy and retain money earned through wage labour, cash cropping or mining benefits.

---

10   In this ritual practice that continued up until about the 1930s, young girls were kept in seclusion in a small hut for a number of months. They were only allowed contact with immediate lineage members who brought food and disposed of any wastes. Typically the girls were being prepared for marriage. Upon their release, a large feast was held for which lineage and clan members brought contributions of shell money that were exchanged for pork. These exchanges initiated new exchange cycles to be executed in future *tolup* feasts.

Plate 4-1: Clan leaders at Lambanam hamlet, Lesel village, 2009. Left to right: Benjamin Rukam, Herman Luak and Joseph Kondiak.

Photograph by the author.

Melanesian Managers

In light of anecdotes and ethnographic records for New Ireland prior to the Second World War (Bell 1934, 1935b; Groves 1934, 1935; Powdermaker 1971), it is reasonable to conjecture that the group surrounding a big-man was stronger and more politically integrated than modern alliances. These big-men were different in so much as they were more active in their recruitment of followers, seizing opportunities to provide security, pay bridewealth and compensation on behalf of others. As we saw in the previous chapter, individual access to cash through wage labour, market sales and *bisnis*, combined with gradual atomisation, meant that most Lihirians no longer solely depended upon big-men or clan affiliation for access to wealth. The situation is somewhat different within 'landowning' clans where wealth is highly concentrated in the hands of those big-men designated as 'block executives' and 'signatories' to clan accounts into which royalty and compensation payments are deposited. Individual clan identity might be self-evident and unassailable from birth, but the clan as a cultural entity (as opposed to bureaucratic political units) could only truly come into existence through the organising efforts of big-men.

Traditionally, men achieved their status and authority through a mixture of fear and love. A big-man's benevolence would endear people to him, and the love he expressed for his followers through his actions on behalf of them was reciprocated through their commitment or love towards him. At the same time, he was respected because people feared his powers and capabilities. Military prowess played a strong part in the construction of a leader. Even if claims to cannibalism cannot be validated (Obeyesekere 2005), Lihirians did fight internally with other clans, and practiced raids on other villages. This is somewhat confirmed by the various stories of 'culture heroes' who brought peace to Lihir, and tales about success in warfare and the competitions between big-men as one sought to avenge the offence of another. While men had to be competent hosts and competitive in exchange, it was imperative that they display their physical strength, usually through fighting, to initiate their status and provide a base for their political careers. Apart from their own autonomous actions, big-men also gained status from *rarhum* feasts held by their clan in their honour.

Assembling the Spirits

Big-men usually held vast amounts of knowledge on matters of sorcery and magic, and would boast of their strong connections with the local spiritual environment. While a man may gain a reputation as a sorcerer, this does not always mean that he will become a significant big-man. On the other hand, big-men without knowledge of sorcery and magic often aligned themselves with those who possessed such knowledge, for their own protection and to boost their own abilities and resources. Big-men who were sorcerers would usually align themselves with similar men. Controlling knowledge was vital to the continuation of their authority, especially given that 'power based on knowledge control may be more subtle than power based on economic control, but this sort of power is also more persuasive in that it involves the communication and validation of cultural categories and truths' (Lindstrom 1984: 299).

Knowledge of the spiritual world was most commonly expressed in the relationship between a big-man and his clan *tandal*, or *masalai*. These are the most frequently mentioned and encountered spiritual beings and remain an important source of power (see Wagner 1986: 102, 107). *Tandal* are simultaneously singular and plural, and incorporeal and manifest in different physical forms. They inhabit prominent physical features around the islands, such as cliffs, caves and large rock formations, sections of reef, creeks and river mouths, points, bays, and even large trees. The connection between *tandal* and certain portions of land is regarded as a sign of ownership by the clan identified with a particular *tandal*. Jessep (1977: 172–202) records similar beliefs in the mainland New Ireland village of Lokon, where clan *masalai* are said to legitimate claims to ownership of certain tracts of land. *Tandal* are also able to transform themselves into animate beings, such as eagles, sharks and snakes, and even incorporeal forms such as strange lights or sounds that can be seen and heard during the night.

Tandal inhabit a central space in the cultural conception of clans. Each clan is said to have their own *tandal* that inhabits particular places and takes the form of certain animate beings that are recognisable to clan members, especially big-men who are expected to maintain close rapport with *tandal* on behalf of the clan. Clan origin stories often feature *tandal* locations as points of emergence. *Tandal* are both benevolent and frightening, and when called upon they inspire awe (particularly when big-men call out the name of their *tandal* during exchanges in final stages of the *tutunkanut* feast). They are metaphorically glossed in Tok Pisin as *strong bilong graun* (the strength of the ground), which reflects the belief that *tandal* provide a kind of invisible moral shield around the islands, protecting them from negative outside forces and thus assisting social harmony. The connection between different *tandal* is crucial to this balance. Negative human activities and changes to the relationship between humans (especially big-men) and their *tandal* weaken the bonds between *tandal*, allowing the further intrusion of socially negative forces. Big-men are morally required to maintain close relationships with their *tandal* for collective benefit.

Paramount Authority

In addition to *tohie*, Lihirians recognised men whose status and authority exceeds all others. These men are referred to as *pukia*, and might be regarded as the biggest of all big-men (see Clay 1977: 22). *Pukia* is the term for a fig tree that has many branches, houses many birds and provides shelter. Its wide-stretching branches are symbolic of the reach and remarkable networks of these men. This image conforms to Hogbin's (1944: 258) description of the North Malaitan 'centre man' who, according to Sahlins (1963: 290), 'connotes a cluster of followers gathered about an influential pivot'. Such men provide advice which is sought by other big-men, and their status is due in part to their ability to unite clans, often across moieties. Some people suggested that these men received deferential treatment, indicating that in the past there were more recognised status differences: certain material and behavioural attributes were said to be the exclusive prerogative of *pukia*.

These men were not 'elected' by any sort of council, but rather as their influence grew they slowly reached a position of paramount authority. Apart from their political prowess and exchange capacities, such men held stores of knowledge on ritual, sorcery and magic, and had the deepest affinity with the spiritual realm. Numerous people mentioned several such men (now deceased) from Masahet and Mahur islands (see Plate 4-2). Others pointed me in the direction of several elderly men described as *pukia*. But in such transformed times, these men hardly exercised a comparable level of influence traditionally characteristic of the title. In addition to being ritual experts, these men now have to be tertiary educated bureaucratic managers, preferably with training as a mining engineer, and for good measure experience in international investment, community development and local governance.

Plate 4-2: Thomas Kut was widely regarded as one of the last great leaders on Mahur Island. When he passed away in August 2009 he was estimated to be at least 100 years old.

Photograph from 2007 courtesy of Chris Ballard.

The House of *Kastom*

The centre piece of Lihirian social, economic and political life is the men's house known locally as *rihri* or in Tok Pisin as the *hausboi*. All females are connected to a men's house through their clan, but it is typically a male domain. Women may enter the men's house which they are connected to in order to bring food, sweep or speak with relatives, but it is rare for them to spend extended time within the enclosure. It is taboo for women to enter the men's house of their husband and his kin.[11] The men's house was more than just the arena for enacting political roles and affirming kinship relations; it legitimated Lihirian culture. It was the locus of social reproduction, containing ancestors and maintaining current generations, and embodying the status of the clan. It was the physical and visual manifestation of social value. Its primary function can be understood in terms of nurturance and social reproduction — the essence of the men's house ethos. It was (and has since become even more so) the most distinctive and recognisable feature of every hamlet.

Theoretically, the men's house is open to all men. Lihirians prided themselves on their ability to provide hospitality for visitors throughout the duration of their stay. At the most basic level, male leadership has always focused on the maintenance of the men's house as a social institution and physical edifice, and the guidance, discipline and nurturance of its younger members. Typically the eldest man within a lineage is regarded as the leader and 'owner' of its men's house, but there is often competition between younger brothers and younger nephews for succession. The men's house is considered inalienable from the clan, which is mainly because it contains deceased clan members. Yet there are cases where it passes from father to son at the expense of the matriline. Typically this occurs when there are no suitable inheritors, the lineage has 'run out', or if particular sons have a strong connection to their father's men's house and can make a special case for inheritance upon his deathbed.

Most Lihirian men's houses are easily recognised by the dry stone wall (*welot*) that encompasses the actual house and the distinctive large Y-shaped entrance (*matanlaklak*) (see Plate 4-3).[12] These carved, upturned tree forks are said to symbolise female legs; entrance into the men's house (symbolic of the womb) emphasises the containment of the matriline. Even when men's houses are in a state of (temporary) discontinued use, the *welot* remains as a reminder of the significance of the site. They are the pride of every hamlet and its associated

---

11   In recent years this rule has been relaxed in the context of funerals. It is not uncommon during burials within the men's house for portions of the *welot* (fence) to be removed to allow grieving widows to enter the rear of the enclosure where their husbands are being placed. In many cases affinally related women choose to remain outside.

12   Some men's houses on the western side of Aniolam are enclosed by wooden fences, which is mainly due to a local shortage of stones in this part of the island.

lineage. Great care is taken to ensure against dilapidation, yet the emphasis is more economic than aesthetic. The men's house is comprised of feasts, as every aspect of its construction must be marked with a pig, which similarly applies to all major maintenance. The construction of the *welot*, the shaping and planting of the *matanlaklak*, the erection of the house frame and the placing of the walls and roof, down to the bed or table-like structure upon which feasting food is placed, all require validation through a small-scale feast of at least one pig within the men's house.

Plate 4-3: Lemdaplo men's house on Mahur Island, 2004.

Photograph by the author.

The necessity to consummate all aspects of the men's house through the consumption of pork also partly accounts for the cases where men's houses are left seemingly unattended and in disuse and poor condition for long periods of time. (Re)construction is more than a matter of building materials and labour supply. Depending of the amount of work required, it may take years to assemble the necessary resources (pigs, garden produce, shell money and cash). The use of permanent materials and new building styles (Plates 4-4 and 4-5) means that men's houses are now symbolic of clan wealth and markers of economic difference. Many of the newer men's houses, especially those in Putput, are reconstructed entirely from permanent materials, with cemented floors and electricity.

Plate 4-4: Distributing the feast food in the men's house, Kunaie village, 2004.

Plate 4-5: John Yaspot's two storey men's house, Laksunkuen, Malie Island, 2004.

Photographs by the author.

Lihirians often describe the men's house as their own 'parliament house'. Historically this is a fitting description; it is where the majority of politics, feasting and exchanges occurred, and where men have made their renown, where they lead, nurture and discipline. This is where men gather to make decisions at the hamlet, clan and village level, and it is where men may simultaneously feel among equals, or realise their position in relation to others. While this description has continued to hold, the political realm — especially the politics of development — has largely shifted outside of the men's house into the domain of the local government, the Church, or institutions and groups connected to the mining company. At the same time, the men's house has become the focus of policy making, identity construction and political authorisation. Often called the *as bilong kastom* (the origin of custom), it has become the objectified stage of customary performance.

The Money of *Kastom*

Lihirian *kastom* explicitly involves the exchange of pigs and shell money (and cash and other commodities) in the series of feasts that surround specific life-cycle events. While trade and exchange historically occurred between the islands of this region, the majority of exchanges were between established partners within Lihir.[13] In the past there was a high level of village endogamy, meaning that most exchange relationships were between residents of a particular area. This seems to increase the impression of parochial exchange patterns, in marked contrast with other notable inter-island exchange regions like the Massim Kula (Seligman 1910; Malinowski 1922; Wiener 1976; Leach and Leach 1983; Macintyre 1983; Munn 1986; Damon 1990). However, internal exchange has always been underpinned by external exchange relations, which is especially the case now that Lihirians rely upon pigs from the neighbouring islands to meet the demands for domestic exchange.

Lihirian shell currency as shown in Plate 4-6 is known as *a le* in the vernacular and more generically as *mis* in New Ireland Tok Pisin, was commonly used throughout the region for a broad range of transactions involving pigs, land, compensation and other ritual purposes associated with the men's house and the life-cycle. In Tok Pisin *mis* is best glossed as *moni bilong kastom* (the money of custom). Ultimately the aim was to have as many high ranking *mis* as possible to purchase large pigs to be used in exchange. *Mis* consists of minute highly polished shell disks (approximately 3mm in diameter and 1mm thickness) strung together in fathoms called *param* in New Ireland Tok Pisin.[14] Commonly known

---

13   See Kaplan (1976) for a reconstruction of these trading routes.
14   Measurements are usually taken as the span from the end of one arm to the chest, thus the distance between two outstretched arms would be two *param*.

types in Lihir include: *pangpang, pabang, emiras, tsien pangpang, tumgiet, met, bobreu, kauas, lolot, tingirip, kolmoni, malyang, zikilde, puas, lerau, zilerau, lemusmus,* and *nuas.*

Plate 4-6: Shell money belonging to the Ilam sub-clan on display in the men's house during the final *tutunkanut* feast, Matakues village, 2008. Decorative adornments (*leku*) are attached to the strands of shell money.

Photograph courtesy of David Haigh.

The colours usually vary between types, from dark red through to bright orange, as well as browns, blacks and whites. The production of *mis* (Plates 4-7 and 4-8) has become the specialty of the smaller islands (known as Ihot). Although a certain amount was made on Aniolam, typically the people of Ihot exchanged *mis* for pigs reared on Aniolam. Filer and Jackson (1989: 63) noted that by the 1980s Ihot people were referring to the production of *mis* as their main business. It has since become an even greater source of income for them, which they now refer to as their 'cash crop'. While this pattern of exchange has long served to supply the people of Ihot with pigs for customary feasting, the local manufacture of shell money is a fairly recent innovation.

Plate 4-7: Gastropod shells (*Patella* sp.) used to make shell money. In Lihir these are known as *gam le*.

Photograph by the author, 2005.

Elderly men have described times when Lihirians did not know the origin of *mis* or the craft of producing it for themselves. Prior to World War Two, Lihirians from Ihot regularly travelled to Tabar with pigs, tobacco and dances, and in return they left with ready made *mis* and certain foods (Groves 1935: 360). Supposedly, the people of Tabar received their *mis* from New Hanover. Consequently, the people of Ihot served as 'middle-men' between Tabar and Lihirians on the western side of Aniolam. The varieties of *mis* originated from different places. *Pabang mis* was acquired via partners from Tabar, *lolot* and *ermiras* were acquired from Kanabu in New Ireland, and shells for making *kauas* came from Lamusmus, Tukul Island, and Djaul Island via Kuat (George and Lewis 1985: 33). Lihirians on the eastern side traded *mis* with Tanga and the Barok and Namorodu area where they acquired pigs for domestic exchange. This exchange pattern where *mis* followed a path from the northwest to the southeast, while pigs travelled in the opposite direction, is captured in a Central New Ireland saying which states that the 'eye' or the source of *mis* is to the northwest and the source of pigs is to the southeast (ibid.).

Plate 4-8: Mathew Bektau of Masahet Island producing shell money, 2007.

Photograph by the author.

A senior man from Malie named Utong once described to me how as a young man he was among the first Lihirians to be shown the art of producing *mis* by people from the Kavieng district. It seems that the skill became more common in Lihir after World War Two. Men worked in secrecy with a great deal of ritual and formality, which undoubtedly increased the value of *mis*. With the gradual loosening of constraints, production now involves teams of women sitting around the hamlet slowly grinding and drilling away at the rough shells. Clay (1986: 193) suggests that, after Independence, Lihirians increased production as the new provincial government actively encouraged *kastom* throughout the province. Although a surplus of *mis* has emerged due to the expansion of production, elderly men continue to hold on to invaluable items which define political rank. However, these items are generally hoarded and have ceased to play an active role in the exchange economy, and as a result, few people know what these look like.

Mis was certainly a genuine shell currency, similar to the Tolai *tabu* (see A.L. Epstein 1963a, 1963b, 1969; T.S. Epstein 1968; Salisbury 1970; Gewertz and Errington 1995). It had more token value than use value, so that it served as a means of exchange — an indirect bridge between goods (Sahlins 1974: 227). At the time of European contact, it probably functioned more as a divisible form of currency than contemporary practices suggest. Lihirians recognise different categories or ranks of *mis*, some of which are used for everyday transactions while others might be regarded as clan heirlooms considered priceless and without exchange value. It is likely that there were more varieties of *mis* in the years prior to Independence. Macintyre (personal communication, August 2003) recalls seeing another type called *ndolar*, which is a thicker, deep red disc and made from a mollusc shell. These have been out of circulation for some time, and even as far back as the 1930s they were regarded as archaic. Many of these valuables have been lost over the years when big-men were buried with their wealth, or else disposed of it prior to their death by either throwing it out to sea or burying it somewhere inland. I found that, while most Lihirians identify *pangpang* as one of the highest ranking *mis*, many are confused or unsure of a definite ranking system for the other types. This differs from Kula ranking systems where people have maintained certainty about rank.

Mis has a definite cash value, which has doubled since the late 1990s, not through scarcity, but through general market inflation. When mining first began, a fine *pangpang mis* was sold for around K40 and *kauas* sold for K30. By 2008, *pangpang* could fetch well over K100 and *kauas* was sold for K70. Access to cash has enabled people to purchase greater supplies of shells from the mainland for local production. These are later used in exchanges or sold to Lihirians or other New Irelanders. Until the 1960s, low-ranking *mis* was used to purchase trade store items, and Ramstad (n.d. 2: 16) notes that *mis* was used interchangeably with cash for minor transactions in the village. Given that it could be easily divided, it is not surprising that Lihirians substituted cash for shell money.

Contemporary pig transactions invariably involve both *mis* and cash payments, and have experienced a similar inflation in price. In the 1930s, a large pig traded with Tabar received between six to ten *mis*. One fathom was worth around five shillings (Groves 1935: 351). In the late 1980s, a large pig may have cost around K100 and 20 fathoms of shell money. In 2004, a medium-sized pig would require at least K1000 and up to five fathoms of *mis*, but it is not unknown for people to spend over K4000 on a single pig. The amount of *mis* exchanged is dependent upon the type used. Occasionally people request a total *mis* payment, but it is unlikely that a Lihirian pig would be purchased entirely with cash.[15] The cash value of pigs and *mis* has become highly important; as we shall see in Chapter 7. Pigs are not just valued in themselves, but also for their total cash value, as increased prestige results from the amount spent for their purchase (and delivery).

Although *mis* has long been used in conjunction with cash and has a monetary value, it primarily derived value as a 'moral' currency. It was prized for 'buying the shame' between newly established in-laws, for uniting clans, and representing the 'honour' or 'social merit' of a particular big-man and his clan or lineage, as an objectified visible validation of the social transactions in which he and his lineage or clan have been involved. It was the big-man who held the clan's store of *mis*, including heirloom and decorative *mis*, and those intended for circulation. The store of *mis* was thus a moral evaluation of both the big-man and his lineage. The same can be said of pigs. Jaw bones (*trias*) from pigs butchered during ceremonial exchanges are often hung up in the rafters of the men's house. These mark the ability to host feasts, provide hospitality, and actively engage in exchanges which create, maintain and confirm social relationships. Together pigs and *mis* are the principal media in Lihirian *kastom*.

Finishing the Dead

In the 1930s, F.L.S. Bell remarked that death is the leitmotif of Tangan culture (Bell 1934: 291). This observation still stands for much of New Ireland. In Lihir, it does not mean that people are suspended in a continual of state of grief and oppression. Rather, ritual and economic life are focused in some way on the various stages of the life-cycle, particularly the series of feasts designed to 'finish the dead', glossed in the vernacular as the *karat* cycle. As in most of New

---

15   This is different from those instances where pigs are bought from people from other islands, who only want cash from Lihirians.

Ireland, Lihirian mortuary feasts have always been more than an opportunity to mourn and memorialise the dead. They contain deep symbolic meaning, and are inextricably woven with political and economic struggles, social reproduction and obligations to both the deceased and their lineage. For this reason any consideration of mortuary ritual must be historically located within the changing political and economic climate of the region.

Although mortuary feasting in New Ireland is normally associated with *malanggan* carvings, Lihirians produced very few of these symbolic icons, although those clans with close ties to Tabar, who are most famous for *malanggan* (see Gunn 1987), were more likely to have used these in successive funeral rites. Generally *malanggan* are produced in Northern and Central New Ireland, and range from elaborate wooden carvings (found in just about any Pacific Islands museum display) to large intricately woven disks (Gunn and Peltier 2006: 41). *Malanggan* serve as effigies of the deceased and are central icons in the process of 'forgetting' and 'finishing'.

In the past, Lihirian mortuary ritual incorporated what might be considered a less conventional form of *malanggan*. During the final *tutunkanut* feast for significant men, the skull of the deceased was painted with lime and red ochre pigments and mounted on the ridge pole of the men's house, or stood on a pole in the enclosure of the men's house to which the deceased belonged. Several skulls could be set up depending upon the scale of the feast.[16] The ultimate purpose of these impressive figures (known as *mormor*) was to remind guests who was being remembered (or forgotten). These figures were treated with magic to enliven them, undoubtedly instilling fear into guests and boosting the prestige of the hosts through the display of their spiritual prowess. The skulls were later interred in local caves and were probably used in different rituals to harness the powers of the deceased. Here the dead (*kanut*) are transformed into an asset for the host clan as they help ensure the feast is made memorable. Mission and government concerns over the supposed sacrilegious disturbance of the dead were the main reasons for the gradual abandonment of this ritual, which was last performed on Mahur in 1987.

In order to illustrate past epistemologies and historical transformations and continuities, it is worth speculating on the connection between these figures and the lack of carved *malanggan* in Lihirian ritual. Throughout New Ireland, different types of *malanggan* clearly served as effigies, icons, vessels of the dead, or even 'pictures' of the deceased at the time of death (Billings 2007: 258), or else as a representation of the ancestors in general, as opposed to specific

---

16   Gunn notes that over-modelled skulls were found throughout Central New Ireland. These were sometimes used on top of carved wooden figures or incorporated into the use of large woven *vavaramalanggan*. There are also instances in Tabar, recorded as late as 1984, where the bones of the deceased were incorporated in *malanggan* rituals (Gunn 1997: 84–85).

individuals (ibid.: 279), stressing collective notions of personhood. Where there is ambiguity or variety in the meaning of various *malanggan* across the region, or perhaps just in the literature, Lihirian *mormor* are quite literally the bones of the deceased. As Hemer notes, this practice was closely associated with a time when Lihirians placed a greater emphasis on the bones still being 'the person' (Hemer 2001: 86). With the introduction of Christianity, Lihirians have come to view the world through a sort of classical Cartesian dualism that separates the spiritual from the physical. Post-mortem emphasis has gradually shifted away from the fate of the body (*kanut*) to the spirit (*a tomber/kanut*). And as I demonstrate in Chapter 7, the emphasis has further shifted towards the politics and economics of mortuary ritual, further displacing the centrality of the deceased. The seamless continuity between the body and the soul, spirit or 'person', may partly explain why Lihirians were not compelled towards 'images' of the dead. Certainly it appears that, where some New Irelanders were more figurative, Lihirians were more literal. Nevertheless, the meaning in these figures, whether *malanggan* or *mormor*, ultimately resides in their use (Gunn 1987: 83). It is to these rituals of social reproduction that I now turn.

Social Reproduction

Although large-scale feasting in New Ireland is primarily motivated by the death or aging of clan members, ethnographic accounts are generally structured by a dualistic approach that emphasises the role and interaction of guests and hosts while largely ignoring the deceased. Hemer (2001) has made similar observations, and she approaches Lihirian mortuary feasting from a tripartite perspective in opposition to the dichotomous structure found in the works of Clay (1986), Wagner (1986), and Foster (1995a). Given that Lihirian mortuary feasts now often occur while the celebrated person is still alive, it is difficult to remove them from the analytical equation, for as we have seen, *kanut* once played a very central role in ritual. In the process of achieving fame and prestige or nurturing the opposite moiety (or at least the assembled guests), individuals and groups are specifically attempting to honour their *kanut*. From a purely political perspective, status is ultimately contingent upon who can show the most respect.

There are over 20 different feasts that mark various stages in the life-cycle and other important events in the lives of Lihirians, all of which involve the exchange and consumption of pigs, garden produce, *mis*, and now cash and commodities. Some of the more significant feasts include: *katipsiasie*, the first pregnancy feast; *kale kiak daldal*, the purification feast that symbolically washes the mothers blood off the new born child; *katipkah*, the first hair cutting feast for the first born son; the *tolup* ritual, which prepared young women for marriage; *minakuetz* and *rapar*, the exchanges involved with marriage; the sacred *rarhum* feast, which honours important men and women within the clan; the *mbiekatip*, which marks the first mourning period; and the *katkatop* (called *pkepke* on Ihot) and *tutunkanut* feasts, known as the *karat* cycle, which are the most elaborate forms of 'forgetting' and 'finishing'. Although an aspiring big-man will utilise all of these opportunities to increase his influence, *rarhum*, *katkatop* and *tutunkanut* are the most important categories of ceremonial feasting.[17] These feasts are vital for establishing the authority of individual big-men and the organisation of supporters and allies of a given lineage. They are highly relevant to the transfer of land rights and lineage leadership. They provide the opportunity for individuals to honour deceased members from their own clan and from others, thus potentially allowing them to claim the wealth and resources of those groups.

Although most Lihirians agree that there are certain stages that must be performed within each type of feast to authenticate the event, there is not always agreement over the correct order for performing these stages. Similarly, while people often speak of an ideal sequence for the three major mortuary feasts, in reality the timing and order of these feasts is subject to prevailing circumstances, clan requirements or individual decisions. Ideally, as a person approached senior age their clan would host a *rarhum* in order to show respect and commemorate their status (effectively the opportunity to attend your own funeral). When this person died there would be burial feasts and the *mbiekatip*, which ends the mourning period. This would be followed by the *katkatop*, which may be held only a few months later, or if the clan is unable to organise the necessary pigs and garden produce, or decides to commemorate several people in the one *katkaptop*, it can be years before this is held. Finally, after a number of years, the clan will stage a *tutunkanut* to 'finish' the deceased, which confirms leadership succession and the inheritance of land and resources. It is likely that there has always been some variation. Since at least the 1960s, big-men from Lataul area have been adding innovative elements as feasting became more competitive. Moreover, people's lives never follow a set trajectory. People often die unexpectedly or at an early age, but certain feasts and obligations still require fulfilment.

---

17   In the past, the *tolup* ritual was considered highly important and was regarded as the *as bilong kastom* (the origin or base of *kastom*) because this was where the shells for future exchange were initially given. These shells were used for purchasing pigs that would be later used in exchange. The first hair cutting ceremony played a similar role, singling out the significance of first-born children, and allowing their entry in the exchange cycle, as *mis* and cash are given to first-born children so that they can purchase pigs.

Rarhum: Exaggerated Respect

Similar to other feasts in the *karat* sequence, the *rarhum* (taboo) feast also provides the opportunity for direct comparison between clans and their big-men. This feast is generally performed when the tooth of a senior man or woman falls out, indicating the demise of the body and their decline towards inevitable death. The tooth may be put on a section of reef or planted at the base of a coconut tree, rendering the area or the tree taboo (*rarhum*) until the appropriate feast has been held to lift the sanctions. [18]

These feasts are intended to celebrate the life of important clan members. Allies and supporters, which include the sons of the honoured person, will contribute pigs, shell money and cash. These cancel any outstanding debts and begin a new round of reciprocity within the *rarhum* cycle. There are strict notions of equivalence within these transactions: pigs and portions of pork exchanged within this feast must be reciprocated exactly at a later *rarhum*. Pork and garden produce are now supplemented by rice and tinned fish which are distributed to the various clan groups who attend. Guests 'pay' for their attendance by contributing shell money and cash to the host lineage. By the 1960s, it was common for Lihirians to use cash in this context. These 'gifts', which are directed by the *rarhum* person (the honoured clan member), are redistributed among those who contributed pigs to the feasts. This wealth is then used for purchasing new pigs that are ultimately destined for use in later *rarhum* feasts.

Besides honouring particular clan members, the *rarhum* feast functions as a mechanism for converting pigs into shell money and cash. Although a group may not necessarily make a 'profit', particularly if they 'spend' a lot in order to honour their member, the prestige gained through their expenditure has greater social currency. Thus *rarhum* provides a material dimension to individual prestige and acts as a vehicle for competition between various social groups. Often the men's house in which the *rarhum* is held is conspicuously marked by a tall woven screen made from coconut fronds placed around the outside of the enclosure, emphasising the element of 'containment' (see Wagner 1986: 153). However, it is said that this feature was introduced in the late colonial period. During this feast, all food within the enclosure of the men's house is *rarhum* (taboo) and must not be taken outside or consumed by women or cross-cousins — separate pigs are cooked for these groups. Only sisters and female cousins (women associated with the relevant men's house) attend this event. The nature of avoidance taboos means that their presence directs male behaviour. In the past, the spatial dimensions of male interaction within the confines of the *rarhum*

---

18   This taboo (area) may be known as a *mok*. Ideally, teeth are thrown onto the reef in front of the men's house (if it is on the coast). This is usually where people dispose of the blood and offal from the pigs consumed during the *rarhum* feast. In the past, it was taboo for males to wash during these feasts. The area of beach used for washing after the feast would become taboo as substances from the *rahrum* pigs were washed into the sea.

feasting area objectified relational status, and the men's house often remained sacred for some time after the feast, so that only big-men were permitted to enter, ultimately distinguishing leaders from the led.

The most important pigs in this feast are the *ber pelkan* and the *balun kale*. The *ber pelkan* is designated for cross-cousins and served outside the confines of the men's house, where the cross-cousins and related women gather. The *balun kale* refers to a particular style of cooking (*kale*) in which food is roasted on the fire. *Balun kale* signifies the loss of teeth that are necessary for eating this type of food, ultimately symbolising the inevitable decline towards death. These pigs are crucial to the moral framework of the *rarhum*, mirroring the Barok equivalent, the *ararum* feast, which 'amounts to the invocation of exaggerated protocols of respect and avoidance' (Wagner 1986: 174).

Tutunkanut: Cooking the Dead

Karat feasts serve similar functions to the *rarhum* feast, except that they are performed on a much larger scale and generate fiercer competition between social groups (see Young 1971). These are the largest and most spectacular feasts that require months — sometimes years — of planning, and vast amounts of pigs, garden produce, shell money and cash. Held over several weeks, with impressive and competitive dancing, and the slaughtering and exchange of what can amount to hundreds of pigs, they are the most stunning display of clan solidarity, wealth and prestige. These feasts are essentially concerned with 'finishing' deceased clan members and transferring their possessions and authority to the next generation. They provide the opportunity to expunge existing debts which the deceased has accrued in both formal exchange and daily forms of nurturance. People can divest their emotional and social bonds and obligations to the dead and create new relationships.

The term *tutunkanut* translates as 'cooking the dead', reflecting earlier mortuary practices that involved the cremation of the body, usually after it had been preserved and stored in the men's houses for some time. Due to the scale and intensity of these feasts, it is common for several clan members to be commemorated in the one event. *Katkaptop* are normally concerned with remembering (*nanse miel*), and carry a sombre emotional tone, while *tutunkanut* is a time to forget and finish (*nanse baliye*), accompanied by a more festive mood. Towards the end of the colonial period, *katkatop* and *tutunkanut* were sometimes collapsed into one big feast. This added greater intensity to the event, and enhanced the prestige of the hosts, but also blurred the distinction between the stages.

The most striking moment during the *tutunkanut* feast occurs when clans present their contributions of cash and shell money to the host clan, whose members use this wealth to purchase more pigs and settle old debts during the next phase of the feast. During this stage, which is known as *roriabalo*, clan leaders will mount a specially prepared round-roofed men's house (*balo*) (Plates 4-9 and 4-10), upon which they publicly present their contribution and indicate their support for the hosts and the next phase of the feast. At some point in the late colonial period, this practice was replaced by one called *roriahat*, in which a specially prepared stage (*hat*) was constructed instead. This 'assistance' is termed either *tele* (help that is given to another person that will later be reciprocated), *yehbi* (meaning 'to put out the fire', a metaphor for the payment of outstanding debts), or *saksak* (which likewise refers to the return of *mis* that has been received previously). The speaker will hold the individual strands of *mis* up for public viewing and then throw them onto the ground one by one to be collected by the receiver. When he dismounts he will remove any *purpur* (Tok Pisin for leaf decorations) to signal that he has rid himself of any burdens (*hevi* in Tok Pisin). Clan leaders who speak at this time will often publicly call on support from their *tandal* as they attempt to display the strength and power of their clan. When *mis* is exchanged in these contexts it must be reciprocated in exact form.

Plate 4-9: Ambrose Silul and kinsmen performing on the *balo*, Matakues village, 2008.

Photograph courtesy of David Haigh.

Plate 4-10: Rongan standing on the *balo* announcing his contribution to the feast, Matakues village, 2008.

Photograph courtesy of David Haigh.

These feasts are ultimately judged according to their memorability. The size of the event, the number of attendants or the range of social groups involved, and the quantity of pigs and other foodstuffs that are distributed and consumed, gauge the status of the deceased and the host clan. Typically, the wealth (which has come to consist of shell money, cash and other material possessions) of the deceased person and their lineage will have been distributed after their death in such a way as to produce a series of pig debts that can now be 'called in' for the final feast. Affiliation with the lineage of the deceased is crucial for receiving what remains of their wealth. Lineages rely upon this affiliation for the successful hosting of feasts, and at this stage it becomes apparent who are group members and who are merely supporters or allies. Ideally, the strength of a lineage is formed by increasing the number of supporters while maintaining the solidarity of the core group members.

Plate 4-11: Women from the Dalawit clan from Mahur Island performing at Kunaie village, 2009.

Photograph courtesy of David Haigh.

In short, *mis* and pigs were the central drivers of this mode of exchange. However, the ultimate purpose of this system was not the production of these two things or the simple exchange of one for the other. Rather, these items were central components in the performance of mortuary rituals that elicited the expression and reproduction of Lihirian values — such as authority, prestige and solidarity — through processes that enabled the reconfirmation and sometimes redefinition of group membership and group property. Although the Lihirian lifeworld was progressively transformed throughout the colonial period and the early years of Independence, in the following chapter we shall see how this world has been very abruptly turned upside down.

5. When Cargo Arrives

In PNG, it is still a moot point whether the economic benefits of large-scale resource extraction balance the tremendous social and economic upheaval typically generated by such activities. If there is a willingness to gamble on the hypothesis that the size of the compensation package will offset any negative impacts from the project, then it is also worth remembering that many of the social divisions caused by mining directly result from an inability to reach consensus on the correct way to distribute this pile of money.

The Bougainville conflict, which began in late 1988 around the Panguna copper mine, remains relevant for understanding the changes occurring in Lihir. When Filer (1990) first wrote about the crisis, he was specifically concerned with its social origins. Working from a rather Durkheimian perspective, he sought to understand the local processes of social disintegration — the particular evolution of social anomie. For a long time Bougainville has been considered a 'special case'. It is for this reason that Filer's hypothesis appears all the more contentious, since he argued that the events on Bougainville were not unique, but instead represented the general tendency for mining projects to have a negative, and potentially explosive, social impact on landowning communities. This was immediately qualified by the important point that various contingent factors — political and economic, historical and geographical — will affect the operation of this tendency. It is unfortunate that history has vindicated Filer's seemingly deterministic supposition. In each of the major mining projects throughout PNG (Ok Tedi, Porgera, Misima and Lihir) the processes of social disintegration have unfolded in unique ways. The 'blow outs' have not all been the same, nor have they necessarily occurred to the same extent, but in each instance local communities have faced common kinds of problems from the delineation of land boundaries, the distribution of benefits, the stratification of society, the inheritance of resources, and the succession to leadership.

Filer's analysis of Bougainville partly informed his predictions on the likely impacts of the Lihir gold mine. In a discussion paper presented to the PNG Department of Environment and Conservation (DEC), Filer distilled these impacts down to their base elements: the stratification effect and the demoralisation effect:

> The stratification effect follows from the probability that different members of the local community will experience the different aspects of the development process in different forms and degrees, and the process as a whole will therefore give rise to new forms of inequality,

> division and conflict within the community. … The demoralisation
> effect follows from the probability that the community as a whole will
> be 'overpowered' by the presence of the project, existing mechanisms of
> social control will be disrupted and devalued, and local people's respect
> for 'custom' will progressively be transformed into a frustrating sense
> of dependency on the project as the source of all their problems and the
> only source of their solution (Filer 1992b: 6).

Since exploration, the stratification of Lihirian society as a result of the unequal
spread of costs and benefits has been the biggest source of internal disruption.
To some extent, this has certainly spilled over into a more general form of
demoralisation and dependency, although this in no way implies that Lihirians
have developed a discourse of inferiority. While the more intangible — but no
less important — impacts upon the values and attitudes of the local people are
apparent, Lihirians have also developed novel political and cultural responses to
the impacts of industrial development. In the following two chapters I will look
at these in more detail. In this chapter I am specifically interested in the wealth–
modernity nexus and the conditions which gave rise to local political elites who
were central to the development of the Lihir Destiny Plan.

Monitoring and Understanding the Impacts

Mining has reproduced the sort of neo-colonial relations that Lihirians hoped
would be lost in the transition to independence. Expatriates occupy managerial
positions, receive the highest wages, accrue the most wealth and possessions,
live comparatively lavish lives, and enjoy greater freedom and higher status
than most Papua New Guineans. On the other hand, Lihirians largely remain
subsistence farmers with village-based lifestyles, have comparatively limited
purchasing power, are constrained by local values, networks and structural
inequalities, and more importantly, only recently began to engage in the sort
of market relations that underpin the existence which most expatriates take
for granted. At one end of the spectrum is the general manager of the mine,
who lives in the biggest house, occupies the most important position within
the company, and receives the most income and perquisites. At the opposite
end are the many aging Lihirians with few means of acquiring cash, who are
reliant upon farming and the generosity of kin for basic items like soap and
clothes, or for luxury food like rice and tinned fish. Lihirians readily endorse
such descriptions, often for various political purposes, but it oversimplifies a set
of exceptionally complex circumstances. Between these two poles lies a range of
social, economic and political positions.

Under the terms of the *Environmental Planning Act 1978*, mining companies operating in PNG were required to develop an Environmental Management and Monitoring Program in which they dealt with the question of social impacts. At the time of the Lihir project negotiations in the early 1990s, there were no detailed guidelines on the manner in which companies should address these impacts. Consequently, there has been little uniformity in the monitoring and mitigation strategies employed across mining operations in PNG — if they have been deployed at all.

In Lihir there has been a gradual development of a more comprehensive monitoring program. As a result we now have a very detailed picture of the social impacts that have occurred since mining commenced. Following the original impact studies by Filer and Jackson (1986, 1989), and Filer's (1992b) discussion paper to the DEC, Filer was later asked by Lihirian landowners to provide them with advice on the sorts of impacts they were likely to experience and mitigation strategies they might pursue. A shorter report, which was funded by the company, was written in Tok Pisin and presented to the landowner association (Filer 1994). After the IBP agreement was signed in 1995, the Export Finance and Insurance Corporation (EFIC), which carried part of the risk insurance for this project, required that an impact monitoring program be designed and conducted according to their specifications. An internal program of reporting was developed, mainly because the company and EFIC thought it was too risky to place the results in the public domain. Martha Macintyre was appointed in 1995 as an external consultant to produce annual social impact reports, and in 1997, she was joined by Simon Foale and the assessment work incorporated a more ecological perspective. Both carried on with this work until 2004, producing valuable documentation of the changing social environment (Macintyre 1996, 1997, 1998a, 1999; Macintyre and Foale 2000, 2001, 2003).

Throughout this time, various other consultants and research students have conducted studies on different aspects of Lihirian social change. The company has also carried out its own internal monitoring program through the Community Liaison Department which includes sections for Lands, Community Relations, Social Development, Cultural Information, Public Relations, Business Development, and Social Impact Monitoring, and which now employs over 60 personnel, the majority of whom are Papua New Guinean. This monitoring work was originally coordinated with Macintyre and Foale's program, and included data collection on health and education levels, and on the local economy, as well as maintenance of the Village Population System database. However, the consistency and quality of this program have been uneven, and from 2004 to 2009 suffered a sharp decline, mainly due to variable commitment, limited resources and staff capacity.

The past inability to properly entrench the 'new competencies' (Burton 1999) for social impact monitoring and community relations work in the Community Liaison Department, relative to the local context, has not only affected the quality of the work that is conducted, but has regularly exposed the operation to increased risk. Notwithstanding the annual social impact assessment reports, there was limited support for monitoring and mitigation work when the operation was under Rio Tinto management. It would seem that lessons were hardly learnt from Bougainville. Since the Rio Tinto agreement was terminated in 2005, Lihir Gold Limited (LGL) has felt the effect of the Rio legacy as they sought to redevelop these programs. However, the company has also faced problems with its own internal restructuring processes and limited institutional capacity within the Community Liaison Department. The mounting community conflicts and political complexities that jeopardise operations on a daily basis have made it painfully apparent that LGL management must commit resources that match the intensity of the social environment. In the meantime, as LGL works out how to address the gap between stated commitments to international standards of best practice and actual performance on the ground, it is likely that the daily efforts of Community Liaison personnel will be continually absorbed by the fight against real and metaphorical fires ignited by a volatile political landscape.

From Gardeners to Rentiers

Most Lihirians can exercise various customary rights over different resources and sections of land for subsistence activities, and many even claim an inalienable connection to some of these places. However, not everyone can claim to be a 'landowner'. While this title appears simple enough, in the Papua New Guinean context it can be rather difficult to precisely define who or what a landowner is (Jorgensen 1997). Landowners are a fabrication of economic necessity and national legislation that simultaneously recognises customary land rights, but also demands a particular conceptual relationship, or separation from the land (see Gudeman 1986: 21). As a result, many Papua New Guineans now find themselves responding to an 'ideology of landownership' (Filer 1997a) that merges customary and legal definitions of ownership with expectations of development. However, the ground upon which they act out this ideology must also be productive, or possess some qualities likely to yield economic profit, or at least be subsumed within the broader area that a company, the government, or some outside agent is seeking to use. As Marx stated in no uncertain terms, 'the worst soil does not pay any ground rent' (Marx 1909: 867) — that is to say, the amount or the possibility of income is overwhelmingly influenced by the location. Consequently, Lihir has been divided by extremely random means: a crude distinction is drawn between landowners and non-landowners, or those who claim ownership over land within the mine lease area, and those who do not. At another level, this is conceptualised in the misleading and antagonising split between the so-called 'affected' and 'non-affected' areas.

In Lihir 'resource rent', or the 'economic surplus' as economists like to refer to it, is paid in various ways through royalties, dividends or equity payments. Other land-related payments made by LGL include compensation, inconvenience payments and occupation fees. [1] Lihirians rarely use the term rent, instead preferring a mixture of overarching expressions like *winmoni* (profit), or *moni bilong graon* (ground money). As we saw in Chapter 2, there was originally some confusion over the nature of these payments, which people were inclined to lump together under the familiar concept of compensation, partly because the local land tenure system does not recognise any category of payments akin to rent. Landowners now distinguish between royalties, compensation (*kompensesen*) and equity (*ekwiti*), but for all intents and purposes the whole range of payments is ultimately perceived as a form of compensation for resource development. Most non-landowners generally conflate all of these terms as a generic income that belongs to landowners. Royalties are part of the larger benefits agreement made with the Lihirian community, but unlike infrastructure, services, housing, business, investment and employment opportunities — the development consonant with the broader definition of compensation in the Integrated Benefits Package (IBP) — not everyone can receive royalties. Even though Lihirians agreed that everyone should benefit from the project, those with landed interests were not convinced that everyone should be awarded equal status.

When the IBP was signed, the lease areas under customary tenure were divided into 140 'blocks' of differing size, each of which was assigned to a 'block executive' nominated by the respective landowning clan. Several of these blocks have been the subject of continuing dispute, causing severe division within villages and between clans. The IBP also provided for payment of various types of compensation to six villages (Putput 1, Putput 2, Kapit, Londolovit, Kunaie, and Zuen) that possess varying amounts of land within the mining lease zone. Not all residents of these villages had equal rights to the land within the lease zones, and in many cases these landowners were not even resident in these villages, instead residing in villages outside the 'affected' area. This is most likely the combined result of village exogamy, virilocal settlement patterns and matrilineal inheritance, and the ability of some men to usurp the titles which

---

1   The range of land-related payments made by LGL for various forms of destruction, land use and agreements include: bush slashing or ground clearing; compensation for loss of land and resources or improvements to land (such as houses and gardens); customary payments; extraction of materials from the ground; loss or damage to hot springs; inconvenience (such as dust, noise and traffic); land use; loss of access to megapodes; relocation; royalties; transport provisions; and trust funds.

might have otherwise been claimed by more junior or less powerful men within their clan. The main compensation agreement recognised 90 block executives with either disputed or undisputed claims over one or more of these 140 blocks. Nearly all of these executives were men, and in the cases where women were given seniority, they remained subject to the demands of their male kin. Within the ranks of these 90 block executives there was further division as 58 of these men had customary claims over land in the Special Mining Lease (SML) zone which entitled them to royalty payments, while the remaining 32 executives only had land claims in the other (less prestigious) lease areas. This group of SML block executives contained an even more 'elite' group of core SML landowners who act as family or clan 'heads' and were responsible (or obliged) to distribute mining wealth throughout their respective clans. Tensions have invariably flared when they have failed to share this wealth according to expectations.

Between 1995 and 1997, some K10.5 million was paid out for various disturbances, including K8.7 million in direct compensation payments — much of which was immediately consumed. If we remember that the average per capita annual income in Lihir prior to 1982 was around K65, lives were literally transformed overnight as payments were made for the loss of ground, trees, gardens, gravesites, men's houses and sacred sites, and the relocation of Putput and Kapit villages. Trips were made to Rabaul, Kavieng, Lae and Port Moresby, with the occasional holiday to Australia. Putput soon began to resemble a poor Queensland suburb as people fenced off their hamlets with cyclone wire, made driveways and enjoyed the evening ambience of newly installed street lights.[2] New four-wheel-drive vehicles were bought and crashed, mountain bikes replaced walking, fancy fake gold wrist watches became coveted items, personal stereos continually disturbed the peace, people wore new clothes, and the face of *kastom* was irrevocably changed as pigs were flown in from as far away as Rabaul and trade store items became standard feasting fare. These Lihirians were going to remake their lives as rich modern villagers with a mastery of technology that would be funded by the mining company. In something resembling a giant potlatch, much of this new wealth was instantly devoured — in some cases never to be seen again — confirming earlier predictions that money would simply blow around like rubbish. [3] At least this is how it appeared to people in the 'non-affected' areas.

Initially, most Lihirians considered that it was right for people relocated from Putput and Kapit to have houses built by the mining company and to be

---

2 In typical Lihirian irony, the installation of street lights in Putput coincided with local claims for compensation for 'light pollution' from the nearby plant site. It is these sorts of contradictions that have been interpreted by other Lihirians as a sign of greediness. Alternatively, they might also represent an ingenious capacity for extracting concessions from the company by whatever means possible.
3 The Mount Kare gold rush was by far the most spectacular instance of short-lived mineral wealth in PNG (Ryan 1991; Vail 1995).

compensated financially for the lack of access to traditional men's house sites, damage to graves, and the loss of gardens and other agricultural resources. However, people not living in the affected areas soon realised that they would have be content to sit on the sidelines. Filer and Mandie-Filer (1998) noted that the split between the 'haves' and the 'have-nots' was particularly acute around the border areas like Putput 2. These people found themselves living close to their relocated relatives and their ostentatious affluence, yet they simultaneously feared the envy and resentment of the non-affected community who tended to 'lump all Putputs together as a bunch of idle, greedy snobs who deserve to have their houses burnt, their cars wrecked, and their daughters raped' (ibid.: 6). This seems excessive unless we remember that all Lihirians looked to the mine as their source of economic salvation. The humiliation that accompanied the unequal distribution of wealth was easily transformed into disillusionment and emotional hostility. The general objection that 'non-affected' Lihirians continued to have to their 'affected' relatives was the latters' refusal to share their newfound wealth.

Plate 5-1: Putput village and the processing plant, 2008.

Photograph courtesy of the LGL archives.

Despite the tendency amongst Lihirians to lump all 'landowners' into the same greedy basket, there are startling differences within this group. Block executives are predominantly male, and are the signatories to the clan accounts into which mining benefits are paid. The original list of 'block executives' and 'block members', as they were then called, was shorn of any matrilineal descendants

— those people theoretically entitled to mining benefits by virtue of clan or lineage membership. It is for good reason that many of these original block executives felt it was unnecessary to include these names. They appealed to customary forms of wealth distribution, which they argued would ensure that the 'right' people received payment. However, listing the names of descendents would have enforced some level of transparency, and would have challenged the male bias by which men maintain control over any money and resources — including land that women are entitled to. In the absence of this measure, disgruntled men and women frequently marched down to see the lands officer in the Community Liaison Department to complain about their exclusion and demand their share of payment. By 2010, many of the land blocks in the SML had been further 'subdivided' between different lineages, so that the number of blocks rose to 106, and the number of block accounts increased to 242.

Neither the company nor the landowner association has ever tried to calculate the exact size of the landowning population, although it is likely that the entire landowning group is less than a quarter of the total population. Both maintain detailed records of payment to individuals, together with lineage, clan and SML block information. Yet neither has shown much interest in defining the size of this group or explicitly naming the rightful recipients. This may reflect the company's disinclination to interfere with what it regards as community or family affairs, and the landowners' continuing belief in the strength of *kastom*.

Only a handful of men fit the classic image that many associate with landowner status. By 2006 a core group of six men had received over K1 million in land-related payments since mining began, and 95 had received between K100 000 and K1 million. There is a larger group of some 214 who over the years had received up to K100 000, and some 1438 people who have received between K10 and K10 000 for various land-related payments. Many minor payments (such as for bush clearing) go to individuals instead of the block account. Generally it is major on-going payments like royalties, and forms of compensation that go to the block account for executive distribution. To further complicate matters, being a member of a particular landowning clan does not necessarily guarantee access to benefits, as payments are generally made at the sub-clan and lineage level.

Classical political economy tended to view rentiers, or landowners, as 'unproductive consumers': rent represented nothing more than a tax on the productive system in order to maintain this group in idleness. They were seen to constitute a class of people who had no active relationship to the production of material commodities; they drew revenue but yielded no productive contribution. It is for good reason that in Lihir this group has become the most envied and despised. Apart from arguments about the loss of land, it may be possible to defend Lihirian landowners on the basis of the contribution they make to the local economy through their ability to spend and consume, or to argue that since no one can charge a price if they do no service, all groups which draw an income must *ipso facto* be productive and their income the measure of their worth to society (Dobb 1972: 67). If 'landowners' hadn't given up their land, then there would be no mine, and hence no development from which everyone can benefit. However, I would be hard pressed to find any non-affected Lihirians who support such rentier apologetics: without the mine there would be no landowners, thus less inequality. It is this kind of abstraction (the justification for rentiers) that Marx was criticising, when he warned against mistaking shadow for substance or appearance for actuality. Crude empiricism will not always provide sufficient insight, but in the case of Lihir, particularly from the perspective of the non-affected people, the reality on the ground makes it increasingly difficult to justify the lifestyle of this emergent class.

Our image of idle landlords who take the fattest lamb from a flock they didn't tend is complicated by the fact that landowners often also want to be labourers, managers or businessmen, or at least secure employment for their children and other relatives. Moreover, landowners face great pressure to plough all forms of wealth back into customary endeavours. Their prodigal spending habits have often been criticised by other Lihirians and New Irelanders as 'wasteful' and 'irrational', bringing deep structural and symbolic changes to ceremonial exchange — a point which I return to in Chapter 7 — but this is also a crucial means of redistributing wealth that stabilises Lihir as it teeters on the edge of a socio-economic fault line. Given that Lihirians regard *kastom* as necessary to the public good, boosting the local 'moral economy', then it could be argued that, in contemporary circumstances, conspicuous consumption through *kastom* is a precondition for the continuity of *kastom* and the balance of society. In this instance, landowner consumption is productive of the collective good. But for those who have become indebted in ways that they simply cannot reciprocate, or who sense that the basis of their traditional authority is now undermined, it is quite likely they would consider the 'moral economy' as a more parochial version of global inequalities where arbitrary differences create and maintain new hierarchies.

Benefit Streams: Seen and Unseen

By 2006, economic stratification was even more pronounced than in previous years. Other villages around Lihir had begun to receive their share of mining benefits and 'development', but the affected areas were still setting the pace. Perhaps the only difference was that non-affected Lihirians were less vocal about their economic status, having grown more accustomed to the new economic order. People were still frustrated and angry about the state of affairs, but many realised that more sweeping economic changes were not about to arrive. As you drove, walked, rode or even paddled by canoe along the coastline in either direction from Londolovit townsite, there was a noticeable decline in the material standards of living with increasing distance from the SML zone. Driving along the road through Kunaie, Londolovit, and Putput, the streets are lined with power poles, there are cars in the driveways, migrant workers and their families rent houses owned by landowners, men stagger around drunk on royalties and wages, cheap plastic toys imported from Asia lie broken and strewn about the place, women and men seem to have brighter and cleaner clothes, and television sets continuously flicker with music video clips, C-grade action films, State of Origin repeats, and the national broadcaster EM TV.

Plate 5-2: Relocation housing in Putput village, 2009.

Photograph by the author.

After ten years of mining activities, a definite pattern for the distribution of mining wealth was established. Wards 1, 2, 3, 8 and 11 have the highest concentration of landowning clans and consistently receive the lion's share of wealth. On average, the remaining wards have annually received well under

K200 000 in land-related payments (see Table 5-1). On closer inspection, geographic distribution is further cross-cut by social divisions and internal differentiations that are not always recognisable. In Ward 2 (Putput 1 and 2), which has received the highest amounts of compensation and royalties, there is a significant difference between different clans, sub-clans and lineages, and men's houses. For instance, in 2006 the Sianus men's house in the Likianba subclan of the Tinetalgo clan received nearly K300 000 in land-related payments. The Unapual and Lepugalgal men's houses of the Likianba subclan of Tinetalgo clan, and the Lakuendon men's house from the Lopitien clan, also received in excess of K100 000. The remaining 34 lineages from a range of clans represented in the villages of Putput 1 and Putput 2 all received less than K100 000, with the majority getting well under K50 000.

Plate 5-3: Comparatively luxurious expatriate company housing, 2009.

Photograph by the author.

Table 5-1: Land-related payments (in PNG kina) by Local Government Ward, 1995–2008.

Ward	1995	1996	1997	1998	1999	2000	2001
				Year			
1	59 878	1 014 923	257 847	272 230	330 467	499 004	1 248 034
2	258 053	2 438 098	1 803 202	597 444	581 598	532 798	1 049 355
3	98 170	367 606	744 407	401 493	558 538	623 527	759 599
4	–	18 741	1900	4856	7818	27 856	142 827
5	–	1631	2616	1825	2429	2734	6035
6	480	7401	4926	–	–	–	6074
7	32 124	116 628	59 463	166 593	209 031	184 258	130 606
8	1099	875 104	337 798	547 562	252 411	302 356	468 168
9	–	17 178	216 448	126 695	191 795	200 363	305 405
10	–	12 235	–	–	–	–	14 293
11	15 531	643 216	53 580	22 498	20 717	61 303	289 224
12	25	118 496	1347	499	505	750	133 330
13	6486	47 748	13 437	39 451	52 936	60 090	46 938
14	1392	20 267	83 313	54 661	191 732	–	1466
15	36 397	12 802	2850	402	–	–	1440
	2002	**2003**	**2004**	**2005**	**2006**	**2007**	**2008**
1	1 124 306	847 510	1 134 712	960 993	1 676 595	1 810 045	2 648 620
2	1 300 251	1 000 084	1 117 065	1 097 306	1 622 202	1 958 227	2 209 172
3	882 705	815 031	1 260 604	760 430	1 269 168	1 599 561	1 600 362
4	301 866	83 179	129 350	124 018	124 413	187 965	154 757
5	8239	6783	–	–	–	–	33 731
6	11 238	565	200	–	–	–	29 721
7	158 358	116 577	92 663	81 537	141 541	225 713	355 543
8	846 590	445 983	579 954	507 252	699 941	975 391	1 088 135
9	456 593	283 696	390 354	175 916	340 030	463 777	463 781
10	25 596	7185	–	2970	18 379	–	882
11	342 020	305 667	395 691	333 932	444 670	1 623 632	692 977
12	54 453	48 865	52 910	54 490	54 858	61 869	58 347
13	80 346	80 063	29 175	27 768	117 352	65 916	77 140
14	720	–	–	14 220	4295	561	–
15	2160	1570	–	45	553	–	160

Note: payments rounded to the nearest kina; Map 2-2 shows wards. Source: LGL company records.

So while ward or village data may indicate a high level of payment, often this is to particular groups and individuals. Such variations reflect the ownership patterns of SML land and the subsequent eligibility of individuals to mining royalties. At the same time, these figures provide no indication of the distribution of wages and earnings across the islands, which definitely counter the imbalances from land-related payments. But in Lihir it is the visual signs of wealth that matter. Wages might afford some comfort, but they are still not high enough to afford a lavish lifestyle.

The Village Development Scheme (VDS) included in the original IBP agreement was designed to provide some form of 'balanced' development and to offset the obvious material inequalities between the relocated villagers in Putput 1 and Kapit and the rest of Lihir. The VDS program initially concentrated on the affected area villages (Putput 1 and 2, Londolovit, Kunaie and Zuen). In 1997, the company began constructing VDS houses around Putput and Londolovit villages. By October 1999, around 80 three-bedroom VDS kit houses had been built in Putput 1 and 2. In late 1999, Putput residents hosted a large feast to acknowledge their new houses. K1.5 million had been allocated annually for this scheme, of which K1 million was used for housing, while K500 000 was allocated for health and education programs. The intention was to try and provide some level of parity between the new relocatees and the other affected areas, to avoid complete social breakdown between people in Putput and their immediate neighbours. Inevitably jealousies, tensions and conflicts still persisted.

In 2000, angry residents from Malie Island stormed the Community Liaison office to protest over the plume in the ocean from mining activities. They demanded that they be recognised as an affected area and be given due compensation. The bush material houses on Malie have since been replaced with permanent structures, many of which contain solar electricity. However, the combination of permanent houses and a rising population, which was already an issue before exploration, has placed even greater pressure upon this tiny island.

It was not until 2001 that there was a more equitable distribution of VDS funding throughout the wider Lihirian community. Each of the 15 wards in Lihir now receives an annual VDS budget based upon their population. Individual wards control their budgets through their Ward Development Committees and Village Planning Committees, which are usually comprised of a group of men hand selected by the local Ward Member.

Throughout Lihir, at least 95 per cent of the VDS budget has gone into housing. In 2005, the total VDS budget had increased to around K3.5 million, which was divided between VDS grants for utility services to the affected areas, and VDS housing grants for 66 houses. By 2008, around 700 houses had been built under the Relocation and VDS projects. Over the years, the VDS has been interpreted as a 'housing fund' for Lihirians. What was originally worded in the IBP agreement as 'assistance to improve housing conditions' is now understood by many Lihirians to mean the provision of a new house for everyone. The ways in which the VDS funds have been used so far have set an unfortunate precedent. Most Lihirians expect to receive their own VDS house in the near future. However, given the costs and the rate of population growth, in 2007 it was estimated by company personnel responsible for the VDS that it would take at least 25 years to build every Lihirian a permanent house. Few people have opted to try and use the funding to improve their existing housing, which in the long run would benefit a greater number of people. Instead, many impatiently wait for what they consider is rightfully theirs.

The frustrations surrounding the slow delivery of VDS houses, coupled with their uneven distribution, have created jealousies and divisions. VDS houses are now flashpoints for continual community conflict; they condense the pressures between nuclear families and the matriline and symbolise material inequality. The ownership and inheritance of permanent structures situated on land held within a system of matrilineal land tenure creates tension between husbands and wives, parents and their children, cross-cousins, and young men and their maternal uncles. Young men who expectantly look to their maternal uncles for support and to inherit resources, such as a VDS house, are feeling increasingly disillusioned as their uncles concentrate upon their own children at the expense of their nieces and nephews. However, when the owners of VDS houses die, it can be particularly difficult for their children to claim ownership of the house if it has been built on their father's land. Often the father's brothers, clan mates and maternal nephews will assume ownership by virtue of the fact that it is located on lineage land. Consequently, some men deliberately build their VDS house on their wife's land to ensure that it can be inherited by their children.

Pseudo Proletariat

Royalties and compensation are more visibly significant, but not all Lihirians have realised that wages and salaries make a far greater contribution to the local economy. In 2005, the company paid K2.7 million in total land-related payments, which barely rates in comparison to nearly K13 million in Lihirian salaries and wages, or the company's reported net profit of US$43.9 million in 2004, US$9.8 million in 2005[4] and US$53.8 million in 2006. On paper, wages and salaries amply balanced the injustice of royalties and compensation, but in reality, they made a minor symbolic dent because few individuals receive wages or salaries which are comparable to the amounts which people imagine that landowners receive. So while frustration was directed at the differential access to resource rents, the uneven spread of employees between villages similarly exacerbated the sense of inequality. Although Lihirians bargained for

---

4   This profit was less than 2004, partly due to the Kapit landslip which claimed the life of two Papua New Guinean employees. The landslip blocked the major road to the township and cut the water supply to the power plant, which caused the loss of almost one month of gold production, reducing gold output by around 100 000 ounces, worth approximately US$50 million in revenues.

'Lihir first' employment policies, wage labourers still comprise a minority of the working age Lihirian population. However, their presence or absence is felt in many ways. Villages closer to the SML zone typically have more workers, while there are fewer in the villages on the southwestern side of Aniolam (see Figure 5-1 and Map 2-2). Similarly, numbers are not evenly spread on the smaller islands: there is a high concentration of employees from Masahet, which partly reflects the higher level of education among Masahet residents.

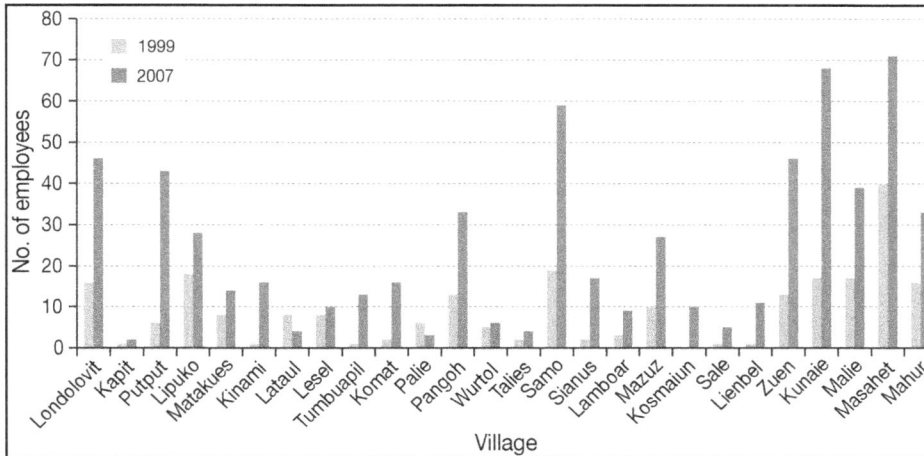

Figure 5-1: Lihirian Lihir Gold Ltd employees, 1999 and 2007.

Source: LGL company records.

Despite these inequalities, wages and salaries constitute the most regular, and the largest, flow of money into the local economy. There was an initial spike during construction, when more than 1000 Lihirians were employed by the company and its contractors, but employment levels then declined. In 2000, the company had around 360 Lihirians on the payroll. In 2002, the company employed 986 staff, of whom 366 (or 38 per cent) were Lihirian, and a total of 90 per cent were from PNG. Of a total 2008 Lihirian population of 14 529 people, 753 Lihirians were directly employed by the company, which equates to roughly 11 per cent of the working age population, comprised of approximately 6720 people. By 2010, there were some 2105 full-time employees, including 697 Lihirians and 1208 nationals, 36 third country nationals, and 164 expatriates.

Even though the mining company is the public face of employment, in reality greater numbers find work with contractor companies and support businesses. In 2004, there were at least 110 companies, organisations and service providers of various sizes (and life-spans) that occupied the surrounding business environment. In 2007 the largest of these were the two Lihirian-owned companies — the National Catering Service (NCS) with some 561 employees, of whom 438 were Lihirian, and Lakaka Civil Construction with 295 employees, including 143 Lihirians.

As in the case of landowner records, LGL has retained limited information on the wider business and employment population. The size of the overall wage labour population is constantly in flux, as local businesses start up and expand, or go bankrupt and close, as individuals move between employment options, and as new migrants seek to create opportunities. If we include people in vital services such as government, health and education, it is quite likely that by 2008 the entire wage-earning population in Lihir — including Lihirians, other Papua New Guineans and expatriates — was in excess of 3500 people. Naturally, within this group there is a diverse range of engagement with the cash economy, and given the nature of fly-in-fly-out arrangements, not all employees are consistently present.

In a kind of Durkheimian organic process, the sheer diversity of businesses, services and economic activity begins to counter the image of the mining company as the totality of the economy. But if the mine were to suddenly close, the effect would be immediate and profound. It is this reality that lies at the core of the dependency syndrome of which the local political elite were becoming cognisant and sought to address through the development of the Lihir Destiny Plan in order to generate alternative self-sustaining economic activity for all.

Objectively, we might speak of an emerging working class: there is a growing section of the community that is dependent upon wages for their lifestyles (not necessarily their existence). However, the specific conditions in Lihir where people have retained ownership of their land, and where wages and other forms of income supplement a subsistence and ceremonial economy, make it structurally impossible for proletarianisation to be complete. Lihirians have embraced opportunities for paid work, somewhat assisted by 'Lihir first' employment policies, but there has never been a strong workers' union movement (see Imbun 2000), which certainly disadvantages non-Lihirian employees. Lihirian mine workers are more likely to unite as Lihirians or landowners than to express solidarity with other national co-workers, which reflects host community expectations and attitudes towards non-Lihirians. Indeed, Lihirians have elaborated the mutual exclusion of landowners and mineworkers to a degree not found in other mine-affected communities in PNG. Lihirian mine workers are remarkably socially invisible and are largely excluded from positions of leadership within the local community. This seems to partly result from the belief that employment is considered a 'right' for all Lihirians, and partly because male prestige and authority are still firmly embedded within the political economy of the men's house — which is augmented by wages and mining benefits. As we shall see in the next chapter, it is also because the real political battles are played out between LMALA, the local-level government, the State and the company for control over larger benefit packages, of which employment is just one aspect.

Although the landowners association (LMALA) officially distances itself from industrial disputes, labour and landowner interests converge in practice when LMALA fights to ensure more working opportunities for Lihirians, or when landowning clans threaten to close the mine to attain business and labour hire contracts. It is the distinctive relationship to the project and the processes of production that provides landowners with the clout to keep them in the 'compensation game' (see Wolfers 1992), and often reduces non-landowners to mere spectators. The first genuine stirrings of a labour organisation surfaced during a relatively peaceful strike that lasted for a week in September 2007. But given that the strike largely arose from tensions surrounding certain expatriate managers, Lihirian and non-Lihirian employees were united in a struggle against perceived White hegemony without any real class consciousness.

In the village context, wage labourers are not always differentiated between types of employment, or between skill and income levels. Workers receive a certain amount of prestige due to their identification with the wealth, power and technological sophistication (the 'modernity') of the companies for whom they work. Pride in this association has morphed into a kind of 'industrial fashion' as work boots, uniform shirts and safety sunglasses are paraded with *laplaps*, popular T-shirts, and cut-off jeans, often making it difficult to distinguish the employed from the hopeful (Macintyre 2008). Recognition is accorded to people in office positions or highly technical roles, but this generally does not increase their status (or their authority) over other workers in the village, nor can it immediately be converted into influence in customary realms — except when it comes to financial contributions.

Workers are contrasted with farmers, but not all workers use their employment as the source of their identity. Often only those with high levels of technical knowledge identify themselves on the basis of their employment positions. General labourers are less likely to use their employment status to gain recognition in the village. Simultaneously, Lihirians continually express resentment and frustration over what they perceive as a 'glass ceiling'. While the company has invested in significant training and localisation programs, these have not met the local demand. Lihirian disappointment indicates the unrealistic expectations they had about the roles they would play in mining operations, and as we shall see, employment opportunities also tend to benefit males over females. Since mining began, Lihirians have predominantly filled trainee, operator, clerk, assistant and apprentice positions. By 2007, there were eight Lihirians in the 43 'senior' positions, one 'superintendent', and only three in the 107 'supervisor' positions. The huge discrepancies between national and expatriate wages continually reinforce this hierarchy. Thus, any prestige gained in the village from being in the company's employ is offset by low status in the employment ladder, accompanied by a familiar feeling of racial denigration.

The growing number of young males disinclined towards garden work, unable to find paid employment, or sometimes simply unwilling to look for it, often expecting employment opportunities to arise by virtue of being Lihirian, might indicate the presence of a growing lumpenproletariat. Disaffected youth with little interest in *kastom* work, no skills, no likely prospects, and a lack of purchasing power, form a disruptive voice within the community, and pose a great risk to operations and community stability. Many are jealous of their paid compatriots, hostile towards 'outsiders' who are seen to be 'stealing' Lihirian jobs, and disillusioned about unfulfilled promises of employment for all Lihirians. Although a high number are regularly moving between short-term work opportunities, the expectation that everyone will find full-time employment remains strong, highlighting chronic dependency and the untenable hopes generated by large-scale resource development. The scarcity of work might add to the prestige attached to employment, but it is also a source of continuing antagonism that reminds people of their unequal access to wealth.

Keeping the Network Out of View

From the outset, mining has generated real and perceived boundaries within Lihir. As people considered the constraints upon their access to wealth and benefits, many began to reassess their relationships, both within Lihir and with people from neighbouring areas. Old relationships were reformulated or 'cut off', new relationships were created internally, and exchange networks spiralled inwards to avoid losing precious resources to non-landowning clans and non-Lihirians.

Within a matter of years people began describing the people of Putput as 'greedy show-offs'. Not only did they resent them for their wealth, but they no longer felt welcome or equal in their hamlets or men's houses. Some relatives invoked sentiments of traditional egalitarianism in an attempt to coerce Putput big-men to distribute their new wealth in a more even-handed manner. In response, these men gradually sought to cut their networks with demanding and less enterprising kin. After all, they argued, they were the ones who 'sacrificed' their land for the project. Following the emergence of internal boundaries and the exclusiveness of Putput, Lihirians without access to royalties or compensation also began reassessing their relationships with non-Lihirians. The latter were becoming an encroachment on other benefits such as employment, or local services like schools, hospitals, the police force and government departments — effectively, their development.

Lihirians who receive royalties, compensation and other economic benefits would probably argue they have not severed their social connections in their attempt to control the flow of wealth (see Strathern 1979). Perhaps in one sense this is true, given that these Lihirians are known to host some of the most lavish customary feasts that continue cycles of reciprocity and increase the chains of indebtedness — transactions that underpin social reproduction and keep the network in view. Even so, they definitely seek novel ways to strategically manage these relationships in order to contain wealth within a limited sphere. Moreover, as money has entered the ceremonial economy it has sustained the cyclical image of exchange and relational continuity. Similarly, emerging forms of possessive individualism reveal the connection between different types of wealth transactions, test the moral grounds of relationships, and shape the ways that networks are perceived and employed. Over time, many Lihirians have refined — and in some cases redefined — the 'rationales of ownership' and the boundaries of inclusion, coupled with a re-categorisation of non-Lihirians in ways that ideologically shift notions of sociality, obligation and reciprocity.

The strategic management of wealth is not uniquely Lihirian, but has been observed in all of PNG's large-scale resource development projects (Filer 1990; Connell and Howitt 1991; Gerritsen and Macintyre 1991; Banks 1996). Senior men have often attempted to bolster their positions of authority, particularly through the control of new wealth, justified through the rhetoric of the benevolent big-man supposedly overseeing the interests of his people. For Lihirians, this has been particularly noticeable in the tension between senior mothers' brothers and expectant younger nephews who look to their uncles for their share of 'clan wealth'. The tendency for senior men to either use this money for themselves, or to distribute wealth along the lines of the nuclear family, has produced a generation of disgruntled young men who feel cheated out of what is 'rightfully' theirs. However, as Filer noted for Bougainville, the contradictions between local custom and the practical distribution of compensation is not just the result of unbridled greed, but perhaps the 'simple absence of a custom which prescribes the proper way to redistribute rent' (Filer 1990: 12). To an extent, the scale and form of wealth associated with large-scale resource development will always overwhelm and undermine customary forms of distribution and inheritance.

Too Many Faces

Migration invariably poses problems for resource development projects, including the added pressure on local resources from the rapid rise in population. As Banks (2006: 263) observed in Porgera, local landowning communities 'shift from being relatively self-contained and known (by the people in them, at least) to being much more diverse and fragmented in terms of people and agendas, and residents experience a loss of control, direction and security in their lives'. During the project negotiation phase, Lihirians were wary of the urban influences which they thought would accompany migration, such as drug use, prostitution, gambling, petty crime, and the importation of weapons, pornography, and more importantly sexually transmitted diseases. These fears prompted local leaders to push for a fly-in-fly-out arrangement. While this has contained the workforce, it has not stopped the flow of people seeking work and other opportunities.

Initially, migrants were arriving on Lihirian shores from within the New Guinea Islands region. In latter years, there has been a sharp rise in the number of people from the mainland, especially the central highlands. Population estimates by the mining company put the migrant numbers at around 5000 people in 2007. However, the data is unreliable, and according to one Community Relations officer, 'the migrant population remains a black hole'. Consequently, Lihirians now attribute new influences to the migrant population, conveniently heaping all responsibility for negative change upon outsiders. These views maintain the polarisation between Lihirians and non-Lihirians. Outsiders are criticised for disturbing the peace and failing to respect local customs and social protocols. Following Koczberski and Curry's (2004: 367) observations in oil palm plantations around PNG, which house large migrant worker populations, these discourses promote and legitimise the political power of the local community and homogenise outsiders.

Nuanced shifts in sociality and fears of outside disturbance or opportunism have been officially manifest in exclusionary policies that seek to remove or refuse entry to all migrants. The Lihir Law and Order Committee, which was established in 2000, and contained members of the local-level government and LMALA, devised a monitoring plan known as the Sengseng Policy, to screen movement in and out of the islands. [5] According to the committee, the greater influx of outsiders is blamed for 'the rapid deterioration of the high moral integrity of the original Lihir society' (NRLLG 2002: 6). Conflating ideas of past and purity, Lihirian leaders were adamant that such malignancy could be stemmed through closed boundaries and limited engagement with outside

---

5   In the vernacular, *sengseng* means to 'move or walk about'.

influences. Future plans have also included an operational unit to carry out the necessary duties of 'border protection', ultimately aimed at installing a visa-like system.

Although some migrants have illegally settled around the mining lease area, many are also accommodated by Lihirians in return for rent or services in kind. The Law and Order Committee took a blanket approach to the situation, theoretically addressing social disturbance and a more deep-seated resentment towards non-Lihirians who use local services and benefit from economic opportunities on the island. However, Lihirian notions of incorporation have made eviction exercises rather difficult. Host communities therefore find themselves in a compromising situation: few have the capacity to manage the migration issues that arise from industrial development, but many also fear future retribution for direct action. Lihir Gold Limited has been reluctant to accept responsibility for the migrant population, claiming that Lihirians have to accept this as part of development. However, LGL is directly implicated. Company managers have continually failed to enforce their own employment policies, which state that non-Lihirians working for LGL and contractor companies must operate on a fly-in-fly-out basis, reside within the company camp, or be provided with accommodation.

It is becoming obvious that economic and population changes associated with large-scale mining have overwhelmed Lihirian notions of hospitality and relatedness. Previous beliefs about the benefits of an expansive and inclusive network have been reconsidered, if not altogether dissolved. Importantly, these changes presuppose a form of possessive individualism: Lihirians have begun to realise that sometimes it is better to limit other people's claims to ownership to certain items and resources, or certain forms of wealth and development. Ultimately, the epistemologies of capitalism, combined with historical experiences and the interpretation of mining through local cosmologies, have become the reference point for considering individual and collective identities. Put differently, the possessive relationship between persons and things in a world of commodity relations has restructured Lihirian relations and social networks.

The Definition of a Lihirian

The localisation of State functions of policing and border control — the repertoire of gate-keeping procedures — is a crucial component in the construction of Lihirian identity (see Foster 1995b: 15–16). Just as the project of nation making is contingent upon establishing an association between people and some definite territory, separating insiders from outsiders, citizens from aliens, Lihirian policies are equally geared towards developing a local ideology of belonging. This process was further developed when local leaders explicitly outlined the 'definition of a Lihirian':

The Lihirian MATRILINEAL system of descent confirms a Lihirian identity. Therefore a Lihirian is a person who is born of a full blood Lihirian mother or born of a 50 percent mother or born of a mother who traced her matrilineal link to a Lihirian clan identity. His/her clan membership is a solid fact of his identity no matter where he/she was born (LMALA n.d.).

Underneath this definition, the authors define a Lihirian according to the following criteria:

- A born Lihirian with full and half blood through the matrilineal descent.

- A person from outside inherits rights by fulfilling customary requirements.

- A person from outside is adopted into a Lihirian family and clan and fulfils customary requirements.

- A person from outside that has lived on Lihir since pre-exploration days and fulfils category 2 and 3 (ibid.).

The paper further explains these categories, with four classes of Lihirian blood identity, and three categories of non-Lihirian. These essentially state that one is either 100 per cent Lihirian — that is, both mother and father are able to trace their descent through their mother's line — or one is a variant of this with declining Lihirian status as the blood runs thinner. Those without full Lihirian blood on their mother's side, or who qualify as either 50 per cent or 25 per cent Lihirian must atone for this through fulfilment of customary obligations: long term residence in Lihir and extended involvement with a clan, in both customary exchange and daily participation in labour and men's house activities. Those with no Lihirian blood in their line of descent must illustrate that for at least ten years prior to mining activities their customary obligations have been exclusively with Lihirian people. The primary purpose of these definitions is to verify the identity of Lihirians and non-Lihirians applying for work, particularly for those positions where Lihirians are awarded first priority. Job applications are screened against the population data base maintained by the Community Liaison Department and the applicant is then assigned an identification category.

In constructing their identity vis-a-vis non-Lihirians and the nation state, Lihirians draw considerably on defensive primordial sentiments. The contiguities of blood, custom, language and shared common descent or ancestry have assumed an ineffable and overwhelming coerciveness for leaders and the community alike. Lihirians regard these ties as more or less immutable aspects of the social person, and as fundamental characteristics of Lihirian identity. It is these points upon which Lihirians articulate a sense of ethnic difference from other New Irelanders and Papua New Guineans, in order to subvert the networks and connections established through historical ties, provincial boundaries and national citizenship. However, the struggle over mining benefits is not always expressed though a neat insider/outsider dichotomy. In 2008, tensions flared between landowning clans that expected to maintain a monopoly on business contracts with the mining company, and other Lihirians seeking similar opportunities. When some landowners began describing other enterprising Lihirians as *ol autsait lain* (outside groups), this further obscured the battle over who were the correct recipients of mining benefits.

Gendered Disparities

It is no exaggeration to state that Lihirian women have borne the brunt of mining activities. However, just as landowners and non-landowners are far from being homogeneous groups, and have experienced the project in a variety of ways, we should be cautious about drawing simple conclusions about the gendered impacts of mining. The women of Lihir have felt the heavy hand of industrial development, but the male world has also been severely destabilised.

Even though Lihir is a matrilineal society, this does not mean that women traditionally commanded exceptional authority, which appears to contrast with other notable matrilineal societies in PNG. Macintyre recalls how she was originally struck by the muted participation of Lihirian women in public life in comparison to communities in Milne Bay (Macintyre 2003: 122). Part of the disparity arises from the relationship between brothers and sisters, rather than between husbands and wives, although this is also a major factor. There may be a close relationship between brothers and sisters, marked by mutual avoidance and respect, and women regularly defer to their brothers and uncles, but in recent times many men have failed to look after the interests of their sisters and nieces. This is only compounded by the lack of recognition of women's status as landowners in a matrilineal society.

Although a woman can achieve a certain standing as the sponsor of a mortuary feast, or if she becomes the owner of a men's house as the last (or the most suitable) leader within a lineage, generally Lihirian women only display their political skills and speak forthrightly when they are in exclusively female gatherings. In short, women in Lihir, particularly younger unmarried women, are accorded a particularly low status. From the beginning of mining exploration, this ensured that women were rather marginal to decision-making processes, despite having their own views, concerns and expectations about the impending changes. Women may have been consulted by well-meaning anthropologists, social workers and community relations staff, but they did not actively participate in project negotiations.

Mining is typically regarded as a masculine domain, with its techno-scientific emphasis and the sheer brute force involved in moving mountains of earth. This view was reinforced throughout the construction period as Lihirian men and women saw relatively few expatriate women with trade skills or in the operation of heavy machinery, ultimately demarcating the mine as a very 'masculine' space. This effectively created a convergence between the 'modernity' and the 'masculinity' of the mine.

Initially, Lihirian men were quite reluctant to allow their women to engage in wage labour. Two main reasons predominated: the fear that women would engage in illicit sexual liaisons once they were away from the constraints of village life; and male objections to female economic autonomy. Men were also adamant that women should not be employed in jobs which men considered to be 'masculine' and therefore modern. In many ways these fears have been confirmed. Women have asserted new economic rights and become significantly involved in the workforce. The issue of declining sexual morality, and the rising number of single mothers, may well be related to increased female mobility and outside influences, but it also results from scores of 'mobile men with money' roaming about the island with nothing better to do than hunt for women and beer. The fly-in-fly-out arrangements might contain the large non-Lihirian male workforce, but this has not stopped migrants settling in the villages close to the mine, nor has it stopped Lihirian women from working in the mining camp, and it has certainly done little to temper the behaviour and attitudes of Lihirian men.

The number of females with paid employment has steadily risen, and women are also engaging in the informal cash sector to a far higher degree through market activities and access to micro-finance schemes. While there is a small but growing number of Lihirian women who earn reasonable salaries as secretaries or other office workers, the majority are employed in menial and subservient positions, working for meagre wages in the camp kitchens and laundries or as cleaners. This type of employment has hardly enhanced their social status, economic position or quality of life. More recently, there have been some notable improvements in recruitment policies aimed at promoting gender equity in core mining activities, but like mining operations in Australia, women remain in the minority (see MCA 2006). Most Lihirian men have only begrudgingly accepted that women will become involved in the workforce, although it would appear that some actively encourage women to seek work, usually in order to access an income. By 2007, there were over 150 Lihirian women employed directly by LGL and more than double this number employed by contractors. [6]

As in many other parts of the world, Lihirian women engaged in full-time employment, or even part-time work, are still expected to fulfil traditional domestic roles like cooking and caring for children. The value placed upon gardening work, combined with the expansion of customary feasting that requires bigger gardens, means that working women must still maintain subsistence and feasting gardens for their family and lineage. The mothers, sisters and daughters of employed women often baby-sit their young children or work in their gardens, for which they gain access to money. However, the relationship is not conceptualised as 'employment', but rather as a means of distributing money to relatives. Macintyre makes the important observation that employment in Lihir has not resulted in major economic change in terms of the dependence of the population on women's labour as subsistence gardeners (Macintyre 2006: 138). Even where households have sufficient income to spend on food, this has not transformed dependence on subsistence to full immersion in the cash economy. Lihirian women — especially those who travel from the outer islands — are now working to a far greater extent (anything up to 12 hours a day), and they remain subsistence producers. As Macintyre aptly stated, 'this is working the double shift with a vengeance' (ibid.: 139).

The greatest impacts upon the lives of Lihirian women have occurred through the disturbances to family life. Women have consistently reported increasing levels of marriage breakdown, domestic violence, rape and child abuse. Many women attribute these changes to male alcohol consumption and the growing presence of outsiders. In the early years of the project, these problems were somewhat confined to the affected villages, but over time they have spread across Lihir. The perception that sexual morality has steadily declined appears to be supported by the number of women who have worked in the camps and given birth to 'illegitimate' children, as well as the number of women who have become pregnant to men who work in the mine. This trend has emerged as a result of women's freedom, but also because employed men have the added attraction of a regular income. Single mothers are frequently stigmatised and treated with contempt by males, who still expect them to remain sexually available. They are often 'punished' and expected to work harder than their married sisters, and are treated with suspicion by other women in the village on the grounds that they are likely to 'steal' their husbands. This partly reflects double sexual standards, but also Catholic theologies that emphasise guilt and shame over mercy and forgiveness.

---

6   See Macintyre 2006 for a wider discussion of both Lihirian and other national women in the LGL workforce.

The cultural disjunctures of the past 20 years are often found in the different ways that males and females experience, perform, engage and instantiate modernity. Lihirian males exemplify many aspects of modernity, and are often encouraged to pursue 'modern' lives, while females are symbolic of tradition and are expected to embody the supposed values of the past (Jolly 1997; Macintyre 2000). Men maintain their monopoly on prestige and the gendered distinction between modernity and tradition, partly through the 'micro-management' of their women. Male moral and cultural status is dependent upon constraining women's sexuality, labour, economic activities, wider cultural knowledge and relationships (Wardlow 2006). Lihirian men commonly appropriate the income of their wives, daughters, nieces and sisters for their own personal consumption (often to purchase beer) or for obtaining pigs or trade store food for *kastom*. This is justified through appeals to traditional forms of wealth management, in which big-men controlled any shell money attained by women and men under their authority. Women are not afforded the same freedom to pursue employment, and their spending habits are extremely circumscribed by domestic responsibilities and male prodigality.

Female marginality is amply reflected in the miserable history of the Petztorme Women's Association (Macintyre 2003). The lack of funding and support in comparison to the male-dominated landowner association has done little to promote the well-being of women or increase gender equity in major negotiations. The women's association has been beset by sectarian politics, which provide an excuse for Lihirian men to argue that women cannot organise themselves and should therefore be excluded from the political process. Of course, the same kind of internal disputes have been present in male dominated institutions, although initially the local-level government and LMALA were better able to conceal these divisions and present a united front to the government and the company. However, even if a Lihirian woman did possess remarkable leadership qualities, the local structures of male political domination would not allow her to exercise those qualities effectively.

Lihirian women might not exercise significant economic or political influence, but since the arrival of the mine there have still been some outstanding improvements to women's lives. Health standards have risen dramatically, assisted by better nutrition, safer housing and water supplies, the improved standard of care provided at the Marahun and Palie health centres, increased access to medical supplies around the islands through village aid posts, women's acceptance of biomedical treatment, and in some cases a reduction in the distances over which women have to carry heavy loads. By 2003, there were noticeable improvements in maternal health care and standards both in absolute terms and relative to national standards (Macintyre and Foale 2003: 78). Access to education and levels of female participation have also improved over the life of the project, and by 2003 were above the national average for rural areas (ibid.: 79). While there are complex reasons for these improvements, related to changing ambitions and shifting cultural attitudes, they have undoubtedly been assisted by a reduction in the necessity for child labour in gardens, coupled with opportunities and desire among women for local employment and involvement in local political organisations.

Even as women appear to be gaining some ground, with improved living conditions and assertion of their rights to greater gender equality through employment or political organisation, husbands routinely resort to physical violence to reinstate the hierarchy that Lihirian men feel is central to their contemporary masculinity. The rising levels of abuse and rape are clearly linked to greater levels of alcohol consumption, but such behaviour also draws upon previous concepts of autonomous big-men and their ability to 'control' men and women within their clan. Existing gender imbalances have been exacerbated in this modern context as traditional sexual divisions are recast to support male dominance in new situations. 'Culture' has triumphed over 'democracy' and new liberal values, as men appeal to custom in ways that are both creative and dogmatic. Macintyre (1998b) has described this as the 'persistence of inequality', readily identifiable in the contrasting ways in which men and women benefit from the mine. However, what appear to be traditional gender-based divisions and injunctions that regulate behaviour superimposed over new contexts may in fact be ways of producing new forms of consciousness or means for expressing discontent or insecurities about changing circumstances.

The association between men and modernity, and their increased freedom to pursue modern forms of prestige, also means that, despite their consistent attempts, senior men are less able to 'manage' younger males. The flip side is that there are more autonomous men who are neither 'managed' nor able to 'manage' other men, so their 'management' efforts are transferred to women with greater force. When senior men try to assert control over their younger male kin, they invariably confront difficulties, particularly as younger working men declare their independence and express frustrations about the ownership of their money, and any constraints on their freedom to decide how and when this might be spent. There are always times when older brothers, fathers and uncles expect that their younger working kin will provide financial support for various occasions and needs. Tensions inevitably flare and end in 'shaming' sessions when young men are unable to produce money, or in some cases, flatly refuse to hand over the product of their labour. Men commonly get drunk in order to precipitate arguments that they would ordinarily be too ashamed to make with their relatives (see Marshall 1979, 1982). Given that these arguments are often excused on the basis of alcohol, inebriation has become an institutionalised means for the expression of normally suppressed aggression.

Bachelors are increasingly eschewing the men's house and opting to build their own place or reside in the family house, especially if their family has received a new relocation or VDS house. This allows them a certain freedom to escape the demands of senior uncles in the men's house. Yet simultaneously, senior men have become conspicuously absent from the men's house as new opportunities and activities divert their attention. These absences undermine generational interdependence and ensure that traditional knowledge is not inherited. On Aniolam, collective garden work is often avoided by young males in favour of waiting for a ride to town to *raun tasol* (just go around) — to look in the stores, meet friends, *painim goap* (find sex), or wait around for working relatives and friends who can buy beer and other items. In comparison to their male counterparts throughout the province, Lihirian men are characterised by their purchasing power and ease of mobility. Increased transport options — by sea, road and air — combined with larger incomes enable Lihirian men to travel greater distances more often, to purchase more pigs, to attend other *kastom* events, to seek women, or simply act as 'local tourists'.

Emphasis has since been placed on men's ability to achieve financial success, maintain control over forms of disposable wealth, achieve political success through economic patronage, amass non-indigenous knowledge through education, training, work experience and contact with people from other areas, display efficiency in the national languages of Tok Pisin and English, and to assert direct influence over the domestic family unit. Male prestige has become increasingly 'decorporatised', and linked to individual interaction with the cash economy. Men no longer gain status from their association with particular big-men, and they are less able to rely on members of their lineage for support and security. Consequently, Lihirian men find themselves in a quandary, being caught between the burdens associated with tradition and modernity. The social pressures of Lihirian society at large dictate that men should be able to perform successfully in both realms simultaneously, for their own benefit and, more importantly, for their families, lineages and clans. However, there is considerable stress involved in meeting the demands of families, *wantoks*, clans, work and modern economic success, ensuring that people meet these obligations and goals with mixed success.

Because men and women are shown to 'perform, consume, appropriate, and inhabit very different facets and locations of modernity' (Wardlow 2002: 147), it is clearly difficult to generalise about the experience of modernity even at the local level. As Wardlow suggests in relation to Knauft's (2002) concept of the 'oxy-modern', 'male' and 'female' may well be separate instances of vernacular modernity. The experience of modernity on Lihir is inflected with masculine characteristics, and masculinity might be seen as modern in relation to femininity, which is associated with *pasin bilong tumbuna* (the ways of the ancestors) or the maintenance of tradition. But this does not mean that all males have equal freedom to be modern in the same ways, that their engagement with modernity does not elicit critique from others, or that they are not all bound by similar ideological constraints from the hyper-reification of customary values. When older males fail to act in ways that accord with these masculine ideals, they are commonly derided as poor community leaders and, in more extreme cases, as morally corrupt. Senior men might be free to pursue individual desires, but they are not exempt from the expectation to personify restraint and dignity.

New Elites — New Inequality

The transformations created through the distinction between landowners and non-landowners, the limited access to employment and the disproportionate opportunities among males and females, have created unprecedented levels of internal stratification that threaten to unravel the very fabric of Lihirian society. Lihir now expresses many of the relationships characteristic of class society. However, in PNG as elsewhere, we cannot solely rely on class analysis as a methodological device through which all social, economic and political phenomena can be interpreted. In some ways, landowners, non-landowners and labourers have begun to imagine themselves as objective groups with their own material interests. Indeed, given that landowner status is somewhat determined by descent, this might even be considered a self-reproducing group. However, overlapping interests such as kinship, moral obligations and an incipient ethnic identity based on control of mine-derived wealth, unites Lihirians and shapes social and economic behaviour, mitigating the social effects of capitalism. Vestiges of traditional society and emotive forms of identification have hardly faded as capitalism developed. These 'non-class' elements continue to exercise considerable influence over the shape of Lihirian history. Indeed, the specific conditions in which capitalism has developed through large-scale resource extraction means that these non-class elements often remain important points of reference for individual and group action. Class is not rendered insignificant, but stratification or inequality can assume other forms.

There are other stratifying elements that are usually identified with 'late capitalism' or even 'post-capitalism', where consumption appears to determine status differentiation in and of itself, independently of the ways that people acquire wealth (Weber 1978: 937; Bourdieu 1984; Appadurai 1986; Miller 1987; Friedman 1994). In Lihir, status differences are closely connected to emergent class structures, but status envy rather than class consciousness has been the phenomenal form of consciousness of inequality. Landowners, and to a lesser extent employees and business owners, enjoy the ability to live sought-after lifestyles, and to a degree they have socio-economically defined identities and interests that are based on the ability to consume in culturally salient ways.

In recent years, a new political elite has emerged. While they are still grounded in LMALA and the local government, they have more power and prestige than other local leaders. The opportunities presented through mining have enabled these men to achieve a level of authority that has decisively shaped the future of Lihir. Some of them have received company-sponsored education and training in fields like economics and law, developed businesses through their monopoly over royalty payments, gained senior positions within the company or the major contractor companies, and assumed prominent positions within LMALA and the local government. Their lifestyles are wholly different from those of grassroots villagers — including many landowners. They enjoy relative economic freedom to pursue an existence that is simply unattainable for most Lihirians.

Several of these men and their families live at Marahun townsite in housing provided by the company, the government and LMALA, while others have remained in their villages, building their own modern houses. Some have spent time in Australia and abroad for work, training or leisure. Their younger children attend the International School while their older children are based in Port Moresby or Australia. They own four-wheel-drive vehicles, work in air-conditioned offices, and can be found playing golf on the weekend or relaxing after work over a few beers at the Social Club. These men are more engaged and comfortable with expatriates and the national political elite than most other Lihirians. These aspiring men have constructed their identities around new influences from Australia and their relationships with expatriate residents. They have learnt that it is possible to embody foreign values and lifestyles and to simultaneously maintain their Lihirian identities and connections at arm's length — though not at too great a distance, given the size of the island and its population. The classic image of the influential and wealthy elite cut off from their grassroots origins has been displaced by images of a global middle class.

However, they maintain a precarious position. Kin relations place huge expectations upon these men to share their wealth and success. They have to negotiate between keeping their extended network open wide enough to ensure their own security, but tight enough to maintain their wealth and membership of the new elite. They may have garnered individual wealth and success, but their support base remains lodged in their lineages and clans, while their authority is built around the twin pillars of the local-level government and LMALA. Both organisations might represent Lihirian interests, but in recent years they have rarely cooperated, and relations have deteriorated through a bitter struggle over control of the benefits package agreed with the mining company.

Although both institutions have been wracked by incapacity, the peculiar leadership qualities of Mark Soipang have enabled him to dominate the political landscape. Much like the leaders of the local government, his authority rests upon his ability to maintain the support of his clan members and to convince the wider population that he can produce the best possible outcomes for Lihir. But he also relies upon the support of the landowning community whose members provide the economic and political leverage for his power, which in turn enables him to contest the government and the company. His enduring authority may have something to do with historical antagonisms towards the national government which have surfaced in a combination of enthusiasm, tension and ambivalence towards the local form of government. It could also stem from a certain moral high ground asserted by the landowning community, and in the supposition that the local government only exists in its current form, or at least with its sizable budget intended for the benefit of all Lihirians, because of the sacrificial generosity of the landowners. It is this very stance which allowed Soipang — through LMALA — to assume a much greater role in the review of the IBP agreement.

The IBP review ultimately consolidated the leadership of Soipang and his cohort. These men were brought together as the Lihir Joint Negotiating Committee. Together they reconstructed historical desires into a new vision — the Lihir Destiny. Without the review process, it is quite possible that Lihir would be on a similar path to Misima Island in Milne Bay Province, where the community was similarly transformed and its members then left to fight among themselves over some residual trust accounts when the local mine closed in 2005. Political factionalism had a role to play, but it could also be more to do with the shorter mine life on Misima. At the point when Lihirian leaders began devising new strategies for maximising benefits and dealing with the quickening disintegration of their society, Misiman leaders were contemplating the prospect of mine closure and an entirely different set of concerns. It is here that we can see how the range of contingent factors specific to each mining project shapes both the nature of its operations and local responses to them. The social impacts in Lihir are broadly similar to those in other projects in PNG, but the combination of historical aspirations and experiences, local political formations, the review of the benefits agreement, and the anticipated longevity of the mine have provided Lihirian leaders with the platform to re-imagine their future and potentially defuse any sociological time bombs. While the realities of the post-mining Lihir Destiny remain to be seen, at this stage it would appear that Lihirian leaders are trying to steer Lihir on an entirely different course. It is to these very particular and rather unusual cultural and political responses that I now turn.

6. Personal Viability and the Lihir Destiny Plan

One of the things they talked about when I went to Lihir, and I heard it, I didn't know whether to laugh or what, they said they are going to make Lihir another New York! (Ambrose Kiapsani, Catholic Bishop of Kavieng Diocese, March 2004).

Though there is no real solution in the Cargo cult — for the Cargo will never come — the ardent wishes and hopes poured into the movement bolster it up and revive it time after time despite failure. And large-scale activities, some of them quite practical, are carried out under the stimulus of these fantastic yearnings (Worsley 1968: 247).

After seven long years, Lihirian leaders, the company and the State finally reached an agreement on the revised Integrated Benefits Package. Several times negotiations reached a stalemate. By 2007, there was considerable pressure from all sides to finalise the review. The company was offering the most attractive benefits package in the history of Papua New Guinean landowner compensation. They were committed to spending K100 million over five years until the date of the next review. This included the entire range of compensation payments and the development of infrastructure and service provision. Lihirian leaders accepted this offer when the company lent their support for the strategic development plans which these leaders had devised throughout the review in response to the changes taking place in Lihir. On 2 April 2007, Prime Minister Sir Michael Somare was flown into Lihir for the signing of the revised IBP that was re-named the Lihir Sustainable Development Plan (LSDP). The Lihir Destiny Plan was finally in place.

The LSDP, or the Lihir Destiny Plan, is more than just a tri-partite compensation agreement between Lihirians, the company and the State. The size of the package might be considered a victory for landowners. But beneath the large sum of money lies a complex and evolving vision of the future. The aspirations and strategies embedded within this agreement represent a significant departure from the ways in which most landowning communities in PNG have approached mining as their road to development and modernity. The myth-dreams and fantasies of a golden city built by the mining company — or at least the provision of lasting services, employment and wealth that the government has failed to deliver — still remain, but they have been turned on their head.

The LSDP has furthered the growing trend towards larger and more detailed compensation packages (Banks 1998; Filer et al. 2000), and it will no doubt ensure that future governments and prospective companies spend even more time at the negotiating table. If it is acknowledged that the Forum process (such as the Lihir Development Forum described in Chapter 2) has helped to develop adversarial relationships between local communities and government agencies as they struggle over the distribution of royalties, equity, special support grant expenditure, and a range of other benefits which can be debited to project revenues (Filer 1997b: 253), then we must consider the review process as a complex extension of this. In much the same way as the original Development Forum, the review process can equally inspire other landowning communities to negotiate new terms and conditions which immediately threaten the deals done to facilitate the development of all existing projects. The irony of both the forum and the review process is that, while they secure greater local participation in negotiating the conditions of resource development, they also undermine the capacity of the government to exercise its legitimate role as an agency of effective development planning in mining project areas.

However, the LSDP is not just about money in the pocket or company-sponsored development. It is the roadmap for the Lihir Destiny, which imagines a new self-sustaining future *enabled* by the mine, but not *dependent* upon a foreign company. It simultaneously pushes for greater immediate participation in the mining process, but pursues a long-term post-mining vision. It is the culmination of past desires, enduring concerns over the social changes from industrial development, the unique circumstances of the IBP review, and an unfolding image of a reconstructed Lihirian society. This was not the first time that Lihirian leaders have tried to transform Lihirian society. In this chapter, I shall trace the historical analogues that informed this process, the ways in which the new political elite reinvented and adapted their own ideas to deal with the prospective and actual social impacts of the mine, and how they have wholeheartedly embraced a perceived modernity which imagines a rather different sort of development.

Society Reform (or Conform?)

The Destiny Plan emerged from the more comprehensive approach to compensation adopted in the IBP, but more specifically its roots lay in the Society Reform Program. First launched in 1993, Society Reform was principally engineered to redirect the social change both realised and anticipated from the mining project. Initial plans for the program were couched in terms of 'rescuing and reviving Lihirian culture' and developing 'fellowship and relationship principles as regards to interactions among fellow Lihirians based on the

Christian principles of love and basic wisdom of the Lihir customary ways' (LMALA 1994: 6). The program's threefold objective aimed to reinvigorate customary activities and relations with Christian moral values, stimulate the local economy through the involvement of all Lihirians in an abstract notion of development, and create social, political, economic and industrial stability. By 1996, the impacts of social change were already being felt throughout the affected areas, and Lihirian leaders — particularly those associated with the landowner association (LMALA) — pushed for the total implementation of the Society Reform Program. This rarely extended beyond rampant preaching by Society Reform leaders with the expectation that people would then reform themselves into customarily grounded moral businessmen.

In light of the Bougainville crisis, Lihirian leaders sought ways to avoid the social conflict occurring at the same time only a few islands away. This was also the first serious attempt to regulate social change, and an early expression of the xenophobia and exclusivity that has characterised Lihirian political ideologies since the formation of the Tuk Kuvul Aisok (TKA). Society Reform resembled an early 'capacity building' exercise. It was proposed in the pre-construction phase that the major political institutions in Lihir would be reorganised according to local concepts of modern bureaucratic hierarchy. These institutions included the Nimamar Development Authority (NDA), a Council of Chiefs, the Lihir Human Development Authority, and the Lihir 'umbrella company' that would stimulate broader economic development. Flow diagrams and images were drawn up to illustrate how these various bodies would work together to execute the 'Lihir Village Development System' and the 'Lihir Master Development Plan' (Glaglas and Soipang 1993). The NDA would generate social and community development with a particular emphasis on law and order, employment and localisation, and community infrastructure.

The Council of Chiefs, which was apparently based on a 'Fiji model' despite a lack of hereditary chieftainship in Lihir (Filer 1997a: 177), would provide a representative body for Lihirian clans, act as the monitoring agency for development, and provide a channel for the dissemination of information throughout the villages. The Lihir Human Development Agency would implement the overall program, promoting and inspiring 'integral human development' and moral regeneration. This would cover areas of religion and worship, customary activities, education and training for leadership, youth, women, family life and health care. Finally, the 'umbrella company' would create a 'business and investment environment' to promote economic development and participation for all Lihirians. This body would work with the NDA to ensure the provision of major infrastructural services. Additionally, Lihir was to be divided into 'zones' that would enable program administration, with traditional leaders or 'chiefs' to represent the clans in these areas. The LMALA leaders believed that this would form the basis for 'orderly administration and operation' across the entire community during and after mining development.

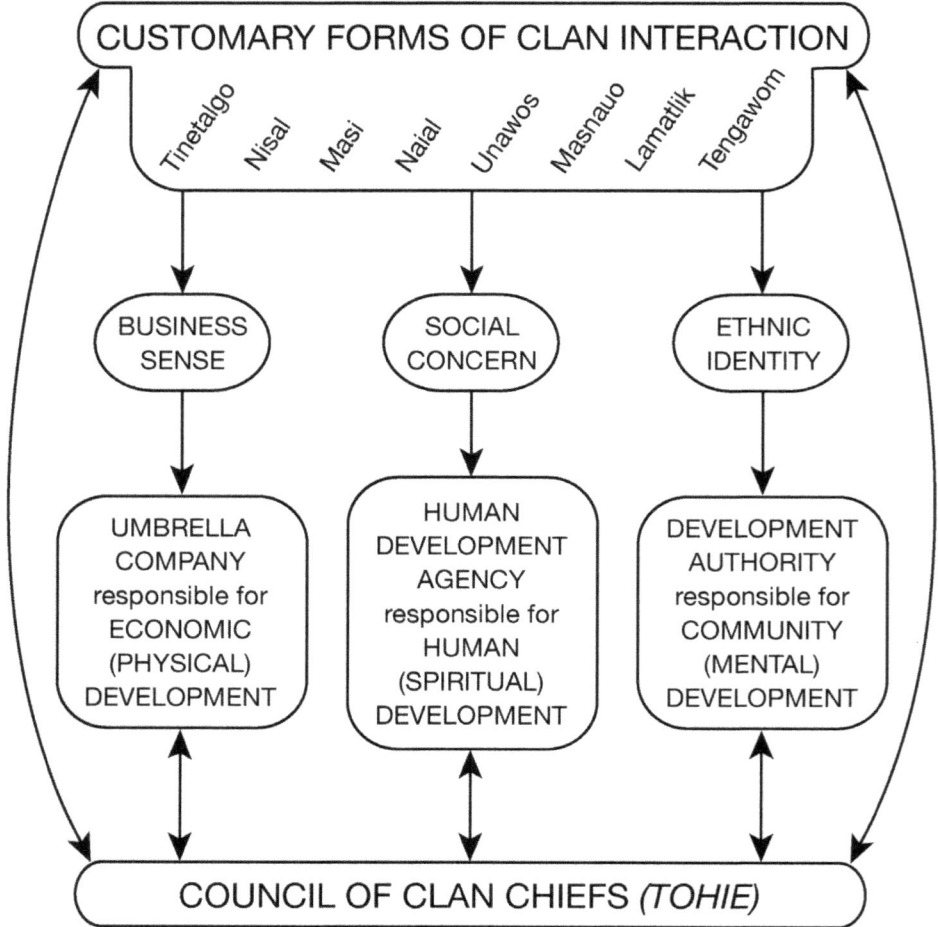

Figure 6-1: Indigenous model of Lihirian society.

Source: Glaglas and Soipang 1993.

Written into these plans was the involvement of Ray Weber, the mining company's first expatriate manager of community relations, specifically for the purpose of providing 'Westerner input' (Glaglas and Soipang 1993: 3). The Society Reform Program was an attempt to escape the cargo cult label that has been applied by outsiders, government authorities and even Lihirians themselves in a process Lindstrom (1995) would refer to as 'auto-cargoism'.[1]

---

1   In addition to previous admonishment from administration patrol officers, Noel Levi from the New Ireland Provincial Government visited Londolovit village in 1980 and told Lihirians to abandon any 'cargo cult-

With its various institutions, Society Reform would show the government that Lihirians could manage the social, economic and political effects of the mine, and that they were no longer cultists but modern bureaucratic leaders able to transform 'traditional Lihir society' into a 'modern Lihir society'.

This was an early attempt to conceptualise institutional regulation of Lihirian life and the rudimentary concept of integration between economics, society and politics, where individual development promotes social development and vice versa. Supposedly, as each person experiences greater personal autonomy and development, coupled with an increased living standard and incorporation into the market, they concurrently support and are supported by traditional kinship and political structures. The early intentions of Society Reform were greater than the rhetoric of 'cultural survival'. Given that the mining project has long been considered as the Lihirian economic millennium — regardless of whether it is viewed through a religious or secular lens — it is obvious that these leaders were not merely concerned with moral decay. What Society Reform envisioned was not just the creation of control mechanisms to shore up Lihirian *kastom* and Christian morals, but a new regulated society, albeit in elementary form, where Western economic institutions are married to traditional social structures to form a distinctly Lihirian modernity. It represented the embryonic stage of Lihirian political and social bureaucratisation and an ongoing attempt to find the right formula for 'balanced' economic development.

Although the program was eventually abandoned in the late 1990s because of its limited results and mismanagement of the company grant, it was certainly not forgotten by Lihirian leaders. It was the genesis of the Destiny Plan. However, it would take the unique series of events that emerged during the review of the IBP to open up a new road for these old ideas.

Reviewing the Future

Once the dust had settled after the initial boom period of mine construction, it was soon obvious that not everyone was going to benefit equally from the project. Society was being 'reformed' but it resembled nothing like what people had imagined. The dense thicket of bureaucracy, chiefs and development agencies built up around the Society Reform Program had completely failed to stem the tide of social decay that was quickly corroding the future viability of Lihirian society.

---

type beliefs'. He also advised them that 'achievement comes only from hard work and co-operation between yourselves' (*Post-Courier*, 27 May 1980, cited in Lindstrom 1995: 43).

After five years of mining, Lihir was suffering heavily from the so-called 'resource curse' and new socio-economic divisions were firmly entrenched. When the IBP review began in 2000, it was primarily seen as an opportunity to reflect upon the first five years of mining, to assess the failures and successes, and to find solutions to new issues and concerns that had arisen. This review process saw the formation of the Lihir Joint Negotiating Committee (LJNC), which included representatives from the LMALA and the Nimamar Rural Local-Level Government, along with a selection of other educated men who formed the new local political elite. As president of the LMALA, Mark Soipang assumed leadership of the LJNC to promote the landowner interests which lay at the heart of the agreement. Together this group was responsible for representing the wider Lihirian community throughout the renegotiation.

It was anticipated that this would be a relatively simple affair to be completed within a year. Members of the LJNC initially saw this as an opportunity to secure more concessions that would supposedly benefit all Lihirians. Their major concern was to redress economic inequality. They had not yet devised plans for post-mining scenarios, particularly for continued service delivery and the maintenance of vital infrastructure and the standard of living associated with industrial employment and higher per capita incomes. Initially, they aimed for increased compensation in the most inclusive sense of the term (see Burton 1997) rather than economic autonomy.

By 2003, the LJNC members admitted they were 'lost' and searching for ways to conclude the review and reach a mutually satisfactory agreement. When they were introduced to the Personal Viability course (often described simply as 'PV') this totally revolutionised their approach to the IBP. As we shall see, the sentiments of PV reflect a mixture of bottom-up development, self-sufficiency, Christian morality, a neo-Protestant work ethic, Western individualism and faith in neo-classical economics. It is a jumble of nationalist and entrepreneurial rhetoric, designed to compel individuals to play their economic role for themselves and their country. Personal Viability soon became the new bible of economic progress and completely transformed the way in which the LJNC approached the future of Lihir. Like fervent missionaries, the LJNC urged Lihirians to abandon their old ways and become 'PV literate'. Personal Viability received mixed reactions throughout Lihir. Some completely rejected it, pitting this road to development against a reified *kastom*. Others embraced the new modernist philosophies and moved from *winmoni* mania to a sure belief that PV would deliver the goods. Before long, PV provided the LJNC with the necessary inspiration for their original Lihir Strategic Development Concept (the precursor to the LSDP, or the Destiny Plan), which for the first time genuinely shifted the emphasis to the long-term post-mine era.

Personally Viable Melanesians

The Personal Viability course was created in the late 1990s by Samuel Tam, a PNG-born Chinese businessman, who argued that PNG needed to become a 'viable' nation.[2] For the past ten years, Tam has been taking active steps to reverse negative trends in PNG with a view to transforming the country from the 'grassroots' up. Where politicians, planners and consultants have searched for policy reform and the right formula for distributing wealth, services and infrastructure, Tam has prescribed strategies that put the onus back on the individual. According to Tam, the apparent development 'failures' over the past 30 years and the palpable decline in national living standards can be attributed to the lack of entrepreneurialism — or 'personal viability'. Thus PV is defined very much in financial terms as part of a broader framework of capitalist economic development to be embraced by each Papua New Guinean.

The PV course is intended to deliver the kind of education that will unleash Papua New Guineans from the constraints that impede economic progress and the improvement of living standards. Tam's vision is that PNG might achieve self-reliance and financial independence by transforming its citizens from grassroots subsistence farmers, bound by a world of *kastom* and parochial economies, into self-sufficient entrepreneurial capitalists, active in a national if not global market.

Born of Chinese parentage in Rabaul, Tam is a reserved man, somewhat suspicious of expatriates, and highly determined to witness change throughout Melanesia. He is tertiary educated, experienced in corporate business and state politics, and affectionately known to PV followers as 'Papa Sam'. His Chinese heritage elicits mixed responses from different groups. For some he represents the new wave of successful Asian entrepreneurs in the Pacific, without the racial baggage carried by former colonial officials, NGO advisers, volunteers, missionaries and emissaries of the Australian government deployed to keep a watchful eye on foreign aid. For many others, however, feelings of ambivalence and hostility towards Asians have coloured their response to PV.[3]

---

2   PV has been endorsed by the national government and various religious organisations that promote it as *the* new answer for Papua New Guineans. The government has regularly used the course for 'capacity building exercises', and there has been a growing interest among community groups looking to advance their own grassroots economic activities (Nalu 2006; Unage 2006). In 2007, Tam took PV to the Solomon Islands, where the Government Caucus reportedly endorsed it as the model for future economic development, and awarded Tam the Cross of the Solomon Islands in recognition of his assistance.

3   To be sure, the growing Asian presence throughout PNG has come under increased local scrutiny, especially for its connections to the logging industry (Crocombe 2007: 64, 134). Asian influence has shifted from small businesses among the 'older Chinese' to corporate investment in the extractive industries, hotels and other areas of commerce, which also seems to be accompanied by greater levels of corruption and organised crime.

Course participants are told various renditions of Tam's rags-to-riches experience, beginning with the death of his father in World War Two, followed by his education in Australia, his early entrepreneurial activities in Port Moresby, his leading involvement in the *Stret Pasin Stoa* scheme, and eventual financial demise that inspired him to develop PV.[4] The story of a businessman of migrant origins, with few familial ties, values wholly derived from modern society, and no customary obligations, has been held up for Melanesians of all classes, educational backgrounds and cultural origins to admire and aspire to. His story has entered what Errington and Gewertz (2004: 15) describe as the 'intersection between different narratives of the desirable and the feasible'; it is presented as the apex of achievement and the new definition of reasonable expectation and accomplishment.[5]

The PV course is part of Tam's wider program for national development to be driven and administered by his Entrepreneurial Development Training Centre (EDTC). So far this program exists in an embryonic form and national implementation has not yet been achieved. The courses are taught by 'trained teachers' certified by Tam's EDTC and working as faithful disciples to spread the good news of his modernist doctrine. Within the EDTC there are various PV courses, ranging from entry or 'village level' to more advanced business courses for the 'PV literate'. His plans also include the establishment of a national *Grasruts Benk* (Grassroots Bank) to act as a microfinance institution for PV members, and the *Grasruts Yuniversiti* (Grassroots University) which will have centres around the country, teaching PV and other related courses, and will act as bureaucratic hubs for the administration of PV programs. These institutions and his administrative system are designed to apply the total PV package throughout the nation and eventually the South Pacific region.

His plans hinge upon a complex grading system for PV followers, whereby 'PV grades' or ranks are achieved by completing various levels of the PV course and measures are made of individual achievement in entrepreneurial endeavours and other facets of people's lives, such as church leadership or 'family management', or the ability to meet customary obligations. The attainment of a bronze, silver or gold ranking objectifies social and economic status and determines how much money PV followers can borrow from the Grassroots Bank. According to Tam, course completion and entrepreneurial achievements will officially determine an individual's 'social class', supposedly motivating further individual economic and political ascension.

---

4   Details of his life history can also be found at the EDTC website (Tam 2010).
5   There are evident correlations between Tam's messages and the prosperity theology found in some of the more charismatic churches throughout PNG.

Two-Week Transformation

I was first introduced to PV in early 2003 when I accompanied a group of men from Kinami to a public forum at which the LJNC was going to explain the progress of the IBP review and its intended development plans. The event was merged with a PV graduation ceremony, designed to demonstrate the new philosophies that would underpin the transformation of Lihir. When we arrived, we found the open-air LMALA community hall in Londolovit townsite bedecked with banners promoting the new road to 'holistic human development'. Some of the banners read:

Lihir Destiny — Self Reliance and Financial Independence

Personal Viability Training Course, Entrepreneurial Development Training Centre Presents: Are You Viable? Holistic Human Development

Lihir Grasruts Benk (Lihir Grassroots Bank), *Benk Bilong Yumi ol Grasruts* (the Bank for the Grassroots People), A Step Towards Financial Independence [this was complete with the image of a pyramid divided into three levels of bronze, silver and gold to represent the various PV rankings that people could achieve with the Grassroots Bank]

Niu Millenuim Gutnius, Grasruts Pawa Mekim Kamap, Universiti PNG, Niupela Rot — Gutpela Sindaun (New Millennium Good News, Developing the Grassroots, University of PNG, New Road — Good Lifestyle)

In Search of a New Direction for PNG — PV Mental power + mental skills + physical power + relevant knowledge + resources = sustainable development (self reliance and financial independence)

Throughout the graduation ceremony, participants shuffled forward to collect their certificates in Holism that certified them as Personally Viable, and then presented testimonies about the amazing personal transformation that they had just undergone. Speaking with excited tones of moral superiority, they attested to the 'truths' of PV and the need to convert from sinful 'handout mentality ways' and step into the bright light of individualism and economic rationalist freedom: *'Yupela yet mas wokim, yupela mas kisim save!'* ('You must do it yourself, you must get the knowledge and skills!'). People were told that 'PV is the Bible in action' and that 'only PV can help our place to develop'. After the forum, which appeared to cause some level of disquiet among the crowd of people anticipating the immediate delivery of benefits and wealth, I spoke to Tam, hoping to learn more about PV. He told me that it was 'just a self-development and personal development course', but it would take at least two weeks to explain everything. I read between the lines and enrolled in the next course.[6]

Personal Viability courses are usually held in village community halls or church buildings to emphasise 'grassroots accessibility'. In 2003, it cost K200 to participate in a basic two-week course. The courses are open to all adults, regardless of their education or work experience, and it is anticipated that children will begin learning PV through the PV Home School Program. Communities can apply to the EDTC for PV trainers to come to their area if there is not already an established EDTC program in the vicinity.

Courses typically begin with a short lecture on the personally viable modern Melanesian based on the following definition:

> PV is the perpetual self-discovery, perpetual re-shaping to realise one's best self, to be the person one could be. It is the sustainable development of human resources with individual skills to be their best. PV involves the emotions, character, personality, deeper layers of thought and action, adaptability, creativeness and vitality. And it involves moral spiritual growth.... ... it is about finding yourself and owning your self (Tam 1997: 11).

Each day begins with prayers and the recital of the PV and national anthems, designed to foster national pride and individual obligation. The course is structured around an ideology of entrepreneurialism (or *bisnis*), and after ten days participants are expected to be familiar with the Entrepreneur's Doctrine (taken from the official creed of the Entrepreneurs Association of America).

Grounded in quasi world systems theory and neoliberal rhetoric, PV aims to create successful entrepreneurs who can reverse the economic trends of the past millennium. Using terms such as core and periphery, marginal and centre, and first and third world, people are encouraged to think of the ways in which Western countries have progressively created conditions of dependence for countries like PNG. Borrowing heavily from Wallerstein's (1974) imagery found in neo-Marxist critiques of the Papua New Guinean post-Independence economy (Amarshi et al. 1979), the idea is to illustrate the global exploitation that keeps countries with people of predominately darker skin on the margins and under the control of the countries in the centre with predominately white skin who are in positions of authority and relative luxury and wealth. Papua New Guineans are presented as cheap labour and nameless peripheral villages are the sites of necessary labour power reproduction for capital intensive centres

---

6   Throughout my fieldwork and subsequent work in Lihir, I have attended five courses, ten graduations, and interviewed over 50 participants.

(Fitzpatrick 1980; Meillassoux 1981). Many participants immediately make the link with the mine and recounted their experiences as 'cargo boys' and 'work boys'. Much of this conversation echoes a more general dissatisfaction with employment opportunities that appear to privilege expatriates over Lihirians.

Personal Viability Anthem

Skul bilong grasrut, universiti	Grassroots school, university
Pawa bilong yumi mekim kamap	Our strength will sustain us
Kamapim pawa bilong manmeri	Sustain the power of the people
Yumi sindaun gut olgeta taim	We will be secure for life
Strong bilong pipel stretim kantri	People power will fix this country
Famili wok bung amamas tru	Families that cooperate will rejoice
Gaden bilong God niu paradais	God's garden is the new paradise
Papua Niugini ples bilong yumi	Papua New Guinea is our place
Sapos mi givim yu pis tete	If I give you fish today
Bel pulap liklik tete tasol	You will only be satisfied today
Nau yumi save pulim pis tu	Now if we know how to catch fish too
Olgeta taim na bel i pulap	We are well-fed all the time
Lo bilong mekim kamap woksmat	The laws that guide efficiency
Lo bilong mani na bisnis tu	The laws of money and business too
Lo bilong kamapim gutpela sindaun	The laws that promote wellbeing
Stretpela pasin stretim yumi	These ways will set us straight
Nau yumi ai op rot i klia	Now we are aware, the way is clear
Rausim ol banis long kalabus	[We can] remove the prison fences
Yumi mas bosim planti risos	We must control lots of resources
Papua Niugini gaden bilong God	Papua New Guinea is God's garden
Yu ouinim yu yet pawa bilong yu	You yourselves own your own power
Yu kepten bilong sip long famili	You are captains of the family ships
Arapela man no inap bosim yu	No one else can control you
Skul bilong grasrut, universiti	Grassroots school, university
Skul bilong grasrut, universiti	Grassroots school, university

Inadvertently demonstrating the internal contradictions of PV ideology, the next stage of the course emphasises that personal failure, poverty and inequality are not the result of a world system that reproduces injustices, but a lack of Personal Viability. On the one hand people are told that their plight can be understood through a structural analysis of global capital flows, and on the other hand the onus is put back on the individual; no longer can people blame the government, isolation or colonial history. This is reinforced through the repetition of popular NGO slogans, such as 'give a man a fish and you feed him today, teach him how to fish and you feed him for a lifetime'. Ultimately participants are taught important lessons in the rhetoric of possessive individualism, particularly as it shadows liberal democracy and promotes the individual as proprietor of the self, who owes nothing to society and is free to act on their own individual conscience. What Tam seeks to foster in Papua New Guineans and the nation as a whole reflects Hobbes' 'self-moving, appetitive possessive individual, and the model of society as a series of market relations between these individuals' (Macpherson 1962: 265).

Practical Lessons in Self-Mastery

For PV apprentices, it is not all pithy self-empowering aphorisms. Students are assessed and expected to meet deadlines for small assignments and group tasks. There are daily exercises in 'grassroots maths' that teach basic requirements for running trade stores and other small businesses. Compared to the time devoted to reciting and rote learning anthems and mantras, a remarkably small proportion of the course is actually dedicated to these pragmatic and useful skills. As a marker of grassroots authenticity, participants are told to do away with ideas of 'laptops and supercomputers', and to start using their own 'neck-top computer'. Drawing upon Papua New Guinean agricultural capacities, Tam encourages people to plant 'money gardens', a PV term that not only refers to cash crops and market produce, but any small entrepreneurial endeavour that 'grows' money. There is a strong emphasis on harvesting 'nature's abundance that is given to Papua New Guineans from God', shifting people's focus from being custodians or stewards of God's creation to exploiters and successful managers.[7]

Within the rubric of personal transformation, participants are taught how to efficiently manage their daily finances and how to say no to the demands and requests of their relatives. For the PV-minded person, it is important to always ask how much money can be made from a particular activity and how personal performance can be improved. Participants are urged to compare their daily routines with the following time chart for the 'average village person':

> **Productive** — gardening, ploughing, weeding, planting; feeding livestock; fishing; building; selling produce; putting the [nuclear] family first.

> **Unproductive** — gossiping; waiting for opportunities or handouts; visiting relatives (*wantok*); sleeping during the day; sitting idle around the village; giving *dinau* [loans].

Participants are warned not to waste their time, their most valuable asset: 'If you cannot control the time you have left in your life you will find it very difficult

---

7   For similar discussions on the ways Melanesians engage with environmental discourse see especially: Van Helden 1998; Kirsch 2004; Macintyre and Foale 2004; West 2006.

to control anything else. … Time is running out quicker than you think!' (Tam 1997: 28–9). In this new order, time is privatised and individuals are responsible for its effective use. These lessons also seek to transform generic notions of village relations by positing the individual (and the nuclear family) as a paramount value. In practice, this tests the moral grounds for relationships, revealing the deep connection between different forms of wealth transaction and individual moral identities. Learning to favour personal ambition over collective stability means that economic imperatives must outweigh social necessities. Thus the highest priority of PV is to quicken the move from develop*man* to development.

Self-Discipline

Personal Viability is divided into 'four growth disciplines': productivity discipline, law of success discipline, economy discipline, and integrity discipline. These disciplines, which can be measured, form the basis of the PV grading system. This is an array of measurements designed to gauge people's individual viability and quantify their ability to 'add value to themselves and commodities' (Tam 1997: 36). By measuring the quantity sold, or the profit achieved, and the rate of expansion and personal progression, 'viable people' can prove that they are constantly 'adding value'. These disciplines specifically target economic output (the productivity discipline), savings and investments (the economy discipline), the ability to meet projected business targets and maintain satisfied customers (the law of success discipline), and finally the ability to fulfil obligations in all areas of life, such as business, family, *kastom*, or church (the integrity discipline). The grading system assumes a lack of motivation: the various ranks are supposed to persuade people to achieve a higher grade that reflects individual earning capacity, credit ratings and supposedly a greater contribution to society. Grading is to be conducted annually by EDTC-accredited grading supervisors. In 2004, PV followers could purchase an official 'EDTC Are You Viable?' badge with the name of the recipient embossed, to be proudly worn to display one's rank and encourage others.

Grading essentially involves verifying the claims of the PV follower: What activities have been completed? What are the annual profit margins? Have expenses, surpluses and savings been recorded, and where are they held? Have sales and productivity reports been produced? And what is the quality of the goods and services being sold? Inspectors should report on the morale of 'The Team' (family or staff who work underneath the graded individual), which means assessing whether younger family members only contribute because they fear retribution, or whether they fully appreciate all the benefits that PV can bring into their lives.

The entire process from reflection to conversion (and public testimony during graduation ceremonies) is significantly theological in tenor. Amidst all of the modernist talk of cultural rupture surrounding PV, these conversion narratives emphasise a clear redemptive strain and an impulse to become clear of past debts and entanglements (the analogues of 'sin'). Personal Viability is presented as the guiding light on the road to development (or re-birth) that will reveal dependency upon the mine and the 'cult of custom' as false gods. However, PV does not demand a total abandonment of *kastom*; rather it calls for regulation and compartmentalisation to curb the ritual proclivities of the develop*man* process.

Plate 6-1: Personal Viability graduation, Lihir, 2007.

Photograph by the author.

The rationalist philosophies in PV resonate with the richly embodied practices of Pentecostals.[8] What emerges from this constant assessment and self-reflection is an extension of the Christian moral ethos of self-examination so that it becomes a natural part of the modernising process for the aspiring subject. Regular

---

8   Contemporary research on Pentecostalism in post-colonial Africa provides a useful point of comparison. In that instance, severing the encumbering ties is seen as a prerequisite for the freedom and new life that salvation offers (see Meyer 1998, 2004; Gifford 2004).

grading increases people's openness to the people and institutions identified with the power and success of the larger world. As Foucault (1984) observed in his review of Baudelaire's (1863) essay, *The Painter of Modern Life*, one of the distinctive features of being a modern subject is the mandate to explore the potential for self-improvement and the ways in which one might become a different person in the future. Modernity is thus characterised by a deliberate mode of relationship to oneself: the intentional attitude of modernity is linked to an indispensable asceticism (Foucault 1984: 41). Personal Viability encourages a heightened self-consciousness and intensifies self-derision in the name of personal development.

Sailing towards Destiny's Island

> We have a ship, the captain and the crew, the fuel, the food, and water supply that we need to make the journey. We also know in which harbour we will berth....The harbour is the Lihir Destiny. The captain is the NRLLG and LMALA through the JNC. The ship is the Lihir Grasruts Pawa Mekim Kamap Ltd. The course is the Lihir Strategic Development Concept (LJNC 2004a: 5).

The ideology of PV soon got a firm grip on the LJNC. Its message contests a history of cargo discourse and activity and the lottery or lucky-strike mentality fostered through mining and the millions of kina that have been devoured through conspicuous consumption with little thought for the future.

Ray Weber's 'Westerner input' into the Society Reform Program had been replaced by Samuel Tam's home-grown input. Tam's disassociation from the mining company, his PNG citizenship, and his Chinese heritage — his non-white non-expatriate non-miner identity — gave him kudos among the LJNC members who wanted to show the company and the State that they had their own (and in their eyes superior) expertise in the matter of business and development. The LJNC members described their early formulations of the Lihir Strategic Development Concept as their version of the Marshall Plan. It was their 'road map' for achieving the Lihir Destiny, which they described as 'financial independence and self reliance that will promote and maintain sustainable development and enable a progressive Christian Lihirian society with a highly educated, healthy and wealthy people' (LJNC 2004a: 16). Although it grew in complexity to encompass personal, social and national development, the focus has always been on some form of *Lihirian* economic progress.

The main elements of the early Destiny Plan can be summarised as follows: the establishment of the Lihir Grasruts Pawa Mekim Kamap Ltd (LGPMKL), the local franchise of Tam's Grasruts Pawa Mekim Kamap (whose title refers to the idea of 'empowering the grassroots'); the total implementation of the PV program, including the PV grading system; the establishment of the Lihir Grasruts Yuniversiti; the establishment of economic projects to create revenue on Lihir (poultry, fishing, timber, garden produce and cash crops); the Lihir Grasruts Benk; land purchases on and off Lihir for future development; business development; and the establishment of a Lihir FM radio station to provide a voice for Lihirians. It was imagined that over a period of years this environment would produce a (micro) 'nation of entrepreneurs'. Similar to Society Reform, the Destiny Plan proposed a total restructuring of Lihirian society to fit a new version of economic development. In both instances, bureaucratisation and increased surveillance were considered crucial components of modernity. However, the fundamental difference was that, for the LJNC, achieving this new state was contingent upon everybody becoming 'PV literate'. According to the popular LJNC adage, PV had to become 'second nature' or the new 'Lihirian psyche'. The underlying assumption was that total commitment to ritualistic performance of PV would somehow affect the overall outcome.

The LGPMKL was the governing body designated to install the Destiny Plan. The name connotes ideas of 'bottom-up' development and implies that local initiatives are more sustainable and capable of reaching better results than 'top-down' planning imposed by the World Bank, the International Monetary Fund, Western NGOs, consultants, the State or the mining company. Yet despite the intentions, the approach was inherently contradictory. The management bias of a small group demonstrates the sort of corporate hierarchy which its members ideologically eschewed. Ultimately the plan was to fuse the LMALA and the local-level government together under this organisational structure to centralise execution and administration of the various projects that fell within the scope of the Destiny Plan.

Political restructuring would bring clan leaders into the development process, with greater connection between the local-level government and the village, and with a focus on strengthening Lihirian *kastom*. Central to this program was the codification and restructuring of customary activity and leadership. The LJNC intended to have these codified 'rules' passed as legislation by the local-level government, and those not practicing *kastom tru* (true custom) were liable to financial penalties. The separation of *kastom* from entrepreneurial activities, 'family needs' and general politics was central to the LJNC's concept of modernity. There would be separate departments in the new governance structure that dealt with different aspects of Lihirian life: health, education, the economy, women, children, youth and the environment. Although many of these offices already existed, either within the local-level government or the mining company, presumably the idea was to consolidate them under centralised LJNC control. Religion, *kalsa* (culture) and sports were lumped together and assigned as one package for monitoring and developing, as activities supposedly extraneous to the economy. If Society Reform lacked direction on how to develop people and society, as individuals and as a collective, PV stipulated a recipe that covered all dimensions — 'physical, mental, emotional, spiritual, and economic' — on the basis of a belief that:

> Holistic Human Development is integral human development and much more. It cannot be defined in one sentence. Holistic here means the whole human being is only a minute but integral part of the universe. What we do affects everything on earth and the universe. It is not enough to develop the whole person although that is essential. The complete human being must also live in harmony with the overall whole, the universe…It is social, business, financial, spiritual, emotional, family, nature, to be your best. Holistic Human Development is everything in life. It is life itself (LJNC 2004a: 54).

Tam's wider program for developing a PV nation involved the monitoring and surveillance of people's activities and progress that would be carried out through the PV grading system — a process of overtly hierarchical observation that normalises judgment. Through consistent monitoring of PV members and their economic activities, recorded in the grading book which they receive upon completion of the course, the LGPMKL would monitor the state of the informal economy and grade individual and collective progress towards 'holistic human development'. The grading system, based on the four 'growth disciplines' (productivity, integrity, economy, and the law of success) would encompass individual lives and rank people according to how successfully they performed certain tasks, ranging from their economic activities through to church and *kastom* obligations. Specifically, these ranks would determine how much money people could borrow from the *Grasruts Benk* in order to advance their entrepreneurial activities. What the LJNC wanted to install was a kind of panoptic discipline that was not specifically designed to produce power for the LJNC, but 'to strengthen the social forces — to increase produce, to develop the economy, spread education, raise the level of public morality; to increase and multiply' (Foucault 1975: 208).

Writing *Kalsa*

In the last 15 years, Lihirians have attended more customary events, eaten more pork, exchanged more *mis*, kina and pigs, and performed more traditional dances than ever before. Yet Lihirians often speak as if *kastom* is in peril, which can leave the outside observer completely flummoxed. Despite the efflorescence of *kastom*, people often commented that '*kastom bilong mipela i bagarap pinis*' ('our custom is totally ruined'), or *'nau mipela no save long mekim kastom trutru'* ('now we don't make true custom'). More *kastom* is not necessarily better, and does not always translate into a feeling that *kastom* is being preserved or strengthened, let alone practiced 'correctly'.

The LJNC was highly attuned to these changes, which it regarded as signs of cultural decadence and decay — symptomatic of 'irrational' economic sensibilities. Central to the Society Reform Program was the need to *'strongim kalsa'* ('strengthen culture'). Given the close association between land tenure and customary exchange, the program would build upon earlier attempts at codifying land rules. No formal cultural preservation program ever emerged, largely reflecting the highly ambivalent attitude towards traditional culture, which local political leaders have often regarded as an impediment to development. However, the PV course seemed to provide the LJNC with a way to place *kastom* within the new economic and political order. The codification of *kastom* was back on the agenda, but this time as a way to preserve the purity of mortuary rituals as well as that of the modern cash economy. The conceptual separation of different daily domains was seen as the basis for cultural continuity and economic development.

Peter Toelinkanut, a former Air Niugini engineer and a member of the LJNC, headed a new Working Cultural Committee to begin this work. The committee carried out its research between 2002 and 2004, spending time in at least one men's house in every ward, discussing and interviewing the men whom they regarded as a reliable source of knowledge. From the outset this was a selective process, designed to produce an even more selective account. Differences between the wards were taken into account, but the intention was to standardise practice across the islands to preserve a particular rendition of Lihirian *kastom*. The booklet produced by the committee begins with the following inscription:

> Major and Minor Customs and Customary Laws: compiled by Lihir Working Cultural Committee, Lihir Island, New Ireland Province, Papua New Guinea.

> Here is a summary of all Major/Minor customs and Customary laws practiced on Lihir Island. From sleepless nights in all fifteen (15) Wards discussing with elders on Lihirian customs and laws practiced in the past we were able to put this together for review by Tumbawin Lam Assembly — Nimamar Rural Local Level Government. The aim of the project is to standardize customary practices and laws on the Island hence preserving and reviving it for use by future generations. Reviving and preserving these customs and laws means reviving and preserving

unity and identity of Lihir which is fast disappearing because of mining operations on the island. When reviewed by Tumbawin Lam Assembly, the project will go before the Legislative Committee who will then draft it as legislation to be effective for all Lihirians to follow (LWCC 2004: 1).

The collection was written in Tok Pisin and laid out like an instruction book for each major feast. It was divided into two sections: 'Major Customs practiced by a majority on the island'; and 'Minor Customs practiced by a minority on the island'.[9] There were limited details on the feasting stages and accompanying land laws, which, if they were followed, were probably more likely to increase disputes. The legalistic approach to *kastom* and penalties for deviation lends itself to analogies with other movements throughout Melanesia that were designed to regulate people's bodily experiences, their orderings of time and space, and their unified and cooperative social relationships (see Schwartz 1962). The codification of *kastom* was more than an indigenous attempt at salvage anthropology; it was also connected to the definition of Lihirian identity, which was in turn driven by concerns over the distribution of mine-related benefits. *Kastom* must therefore be considered as a particular aspect of social change and as an instrument in political and ideological struggles that Lihirians waged against their neighbours, the nation, and the mining company, and also amongst themselves.

Towards the end of 2004, the LJNC included a Kastom Lidas Komiti (KLK) in their IBP proposal and claimed that the entire 'Destiny Plan' now depended upon the implementation of this new political structure. The KLK would represent all Lihirians. A leader would be selected from every clan in each ward, which theoretically meant that everyone would be represented by a customary leader. One of the main functions of the KLK was the dissemination of information from the LGPMKL to the men's house, supposedly linking everyone to the developmental process. These leaders would revive customary teaching and strengthen respect for men's house ethos. Ideally they would become agents of modernity who taught PV by day and *kastom* by night in men's houses that were converted into multi-disciplinary learning centres.

Unrest

The proposed Destiny Plan was quickly generating a great deal of heat, which arose from the growing social isolation and political dominance of the LJNC, confusion surrounding their plans, and emerging frustrations with the PV movement. The LJNC referred to itself as a 'master-mind organisation' — PV

---

9   In some instances Lihirians refer to the entire group of islands as simply 'Lihir Island'.

jargon that reflected the status which the members believed was conferred upon them by the diversity, educational background, and experience of its members. They were a self-styled group of elite Lihirian men who considered themselves to be the vanguards of social change. LJNC members purposely chose not to work for the company. This was partly due to their oppositional stance towards the company, but also because wage labour was seen as a sign of dependency and contradictory to the entrepreneurial emphasis.

Their elaborate plans, which centred upon the institutionalisation of PV, were largely being devised in isolation from other representative groups. They often stated that they would only talk with people who were 'PV literate' and equipped to understand their plans. Relations with the local-level government deteriorated, so by 2004 the LJNC existed as a separate political entity. Not surprisingly, Lihirian women were completely excluded from this process, which eventually prompted the women's association to launch a protest march through Londolovit townsite to present a petition to the government and the company to demand a greater level of inclusion.

The Destiny Plan was not a community movement, nor did it have full community, government or company support. This meant that people who were not directly involved with the LJNC often found it very difficult to understand its plans and vision. The LJNC supposedly represented community interests, but it was in constant tension with other organisations and the broader community. Although Mark Soipang gained his authority from his position in the LMALA, which was fundamental to the LJNC's control over the revision of the IBP, Soipang had an ambiguous and strained relationship with the landowning community. Not everyone supported his leadership, and as long as they were living off mining benefits, many struggled to find PV relevant. They were more concerned with the compensation details in this agreement than with ideologies of cultural conversion.

When Bill Gates Comes to a *Hausboi* Near You

In order to address rising discontent, the LJNC embarked on a community awareness campaign. People were told they could get on board the boat sailing to the new Lihirian future, or else sit on the shore waiting for the 'white ships full of cargo that will never arrive'. Through the national implementation of the PV course, with Lihir at the head of it, the LJNC depicted itself as the captain at the helm ready to sail PNG through its financial tempest. These meetings were normally held on Sunday mornings after church in public areas in a village. On one occasion, the LJNC held a special meeting in Lataul, in the men's house belonging to the father of one of the LJNC members. This area had remained a stronghold of the Nimamar Association and developed a particular antagonism towards the LJNC and the PV movement.

That evening, when I arrived at the men's house, I wasn't surprised to hear a portable generator whining away to provide power to illuminate the setting. What I did not expect to find was a PowerPoint presentation of the Destiny Plan being beamed onto the inside wall of the men's house. It seemed a slight contradiction to present the future for 'neck-top computer' specialists on a brand new laptop computer. Through a thick fog of smoke the apocalyptic words 'Lihir Destiny' magically hovered above the heads of the older men; I waited for trumpets and horsemen. The presentation began with pre 9/11 pictures of New York with the twin towers still standing as proud monuments to man's achievement. These were followed by more images of Singapore, Kuala Lumpur and Sydney, beamed through the haze onto the dirty inside wall. Gasps of amazement were directed at both the images and the technology. If the LJNC couldn't convince the audience of the rationality and necessity of its plan, then it could dazzle them with bright lights, or impress them through detailed descriptions and complex flow charts which explained how this city would function and the role that each individual would play. These education programs had a similar format to Society Reform presentations and many of the public awareness campaigns by the company on issues such as health and environmental impacts, which usually involved complicated technical and scientific explanations that were rarely understood, but were presumed to be sufficient motivation for people to adjust their lifestyles.

The idea of Lihir becoming a *siti* (city), a generic pie in the sky that promises greater purchasing power and an inverted world order, is a recurrent theme in Lihirian history. This *driman* (dream) has evolved from the deceptively simple *kago* (cargo) to images of a thriving metropolis. Lihirian urban dreams were previously based on their knowledge of Kavieng, a small and sleepy provincial centre on the northwestern tip of New Ireland. Access to media and the opportunity to travel overseas has increased local awareness. However, as the Bishop's mocking remark indicates, the problem was not to imagine how a Lihirian city might look. People had been doing that for at least the past 30 years, and most people already imagined that their urban role would be characterised by consumption, not labour or production. Instead, people were concerned about the overriding emphasis on PV as the means for constituting this Arcadia. Not only did it appear to contradict earlier prophesies, which promised that the ancestors would deliver the desired change, but people were becoming rather wary of the new sociality encouraged by PV.

Personal Viability in Practice

Many Lihirians, especially those without access to royalties and compensation payments, initially thought that PV was the key to wealth accumulation. Large numbers eagerly enrolled in their nearest course, believing that they could achieve their dreams through this 'home-grown' approach. Indeed, this is why it was appropriated by the LJNC — to demonstrate to the company and the government that LJNC members understood the 'White man's secret' and had their own (alter)native answers to the development riddle. Yet large sections of the community were growing disillusioned with PV, which strongly influenced the way in which they interpreted the LJNC and its Destiny Plan. The skewed mining economy and inflated community expectations deterred many people from seriously committing themselves to entrepreneurial endeavours. The LJNC and some company managers were inclined to regard this as a simple extension of the so-called 'cargo cult mentality', which would explain why they failed to recognise that the heightened sense of traditionalism was the greater barrier.

In response to the sense of cultural rupture, many Lihirians have reified their cultural values in a form of hyper-traditionalism which has developed in tandem with the expansion of feasting and exchange through access to new forms of wealth. In the next chapter I shall examine this pehonomenon in more detail, but for the moment it is sufficient to note that, for many Lihirians, *kastom* denotes a form of sociality that supposedly differs from that associated with the cash economy. More often than not, people find that their attempts to conduct business within the village setting, or their strategies to 'get ahead', are constantly compromised by these competing values. The followers of PV like to imagine that their economic activities are set apart from the petty market sales of the average villager, not least of all because they aim to advance beyond ad hoc sales to a regular income supported by a growing 'clientele base'. More importantly, they often state that there will be no bartering, exchange, credit or favours, regardless of kinship and *kastom* obligations. PV projects are ultimately supposed to operate separately from the gift economy. Although the integrity discipline purports to measure individual performance in *kastom* to the extent that such transactions reflect a form of economic commitment, PV implicitly encourages less — or at least more rationalised — involvement in *kastom*, which appears to sustain the sharp ideological division between notions of *kastom* and *bisnis*.

One of the ways in which people are encouraged to perform their viability is by organising their family according to a corporate structure in which each member continually proves his or her productivity by contributing income towards the family's living expenses. Weekly meetings should be held and records must be kept to ensure that productivity levels increase. It is the nuclear family, not the matriline, which is the base model for micro-collective enterprise. This will supposedly instil the PV mentality at the household level, eventually reinforced by a community of like-minded families.

The next level is the formation of village-based PV clubs, or collectives that pool resources, finances and labour for individual and group projects. Some function as a micro-finance resource, providing small interest-free loans to members (anything up to K200) to start another 'money garden' project, pay school fees, or deal with emergencies. From Tam's perspective, Papua New Guineans left to their own devices inevitably fall by the wayside of PV. The clubs provide support and encouragement for floundering entrepreneurs. Modelled on corporate organisations, clubs elect a president and various office holders, with committees established for different projects. Ideally, there should be a club in each village and a club representative in each local government ward. The latter's duty is to make sure that clubs function effectively, help to recruit new followers, and guard against declining enthusiasm. But in Lihir it has often proven difficult to enact these ideal strategies.

PV clubs cut across clan and family ties, and exclude anyone that has not taken the course and proven their 'PV literacy', which in the village context has increasingly come to mean the ability to converse using PV idioms, rather than demonstrate any recognisable form of entrepreneurial initiative. The use of commonly understood terms and phrases exclusive to PV followers generates a sense of ethos among club members. This is important because their petty market sales are sometimes hardly distinguishable, in practice, from those of non-PV followers.

PV urges people to compartmentalise their daily life in terms of a perceived distinction between socio-economic spheres. In reality, it often proves impossible to pry apart the tangled relationship between *kastom* and *bisnis* through ritual adherence to PV, even though it promises some level of financial autonomy from local webs of kinship and exchange. It is this tension that causes people to reject PV, claiming that '*PV em i no olsem kastom bilong yumi*' ('PV is not like our custom'), or '*em i no pasin bilong Lihir*' ('this is not the Lihirian way'). This new resistance is couched in the discourse of *kastom*, but it also comes from a realisation that any 'secrets' which people are learning from PV are difficult to perform, and usually contradict cultural dispositions, values and expectations. Many try to establish their 'money garden' only to find their efforts strangled by tangled kinship roots. Some people might consider themselves as potential businessmen, but they are often the same people who disparage those who are successful at the expense of social relations. Such reactions capture the contradictions and paradoxes in the Destiny Plan and broader Lihirian desire: Lihirians want to learn how to produce money for themselves and gain financial autonomy, but not necessarily at the cost of sociality.

State of Exclusion

It was imagined that PV would help Lihirians to create their own 'nation of shopkeepers and managers'. The LJNC members were in a unique situation. Their access to resources through the mining project, their commitment to PV, and their involvement in Tam's nationwide program converted earlier micro-nationalist sentiments into actual plans for *micro-nation making*. The LJNC had not merely expressed a typical micro-nationalist desire for greater autonomy (not necessarily full-scale secession), but went to great lengths to create a blueprint for a micro-nation with its own laws, economy and leaders, that would not only show the way, but might eventually fulfil prophesies of becoming the *las kantri* (last country). The committee's outlook was ultimately insular, and its members relied on Tam for instruction, inspiration and engagement with the nation through his program. There was an intrinsic tension between wanting 'bottom-up' development and having an inherently 'top-down' approach. The plan was classically neoliberal, based on a reverse trickledown effect through which the rest of the nation would develop as Lihir led the way, after first transforming Lihirian leaders through a process which the LJNC called 'mind conditioning'. The proposal was for more than a mere agreement between the various stakeholders in the mining project, or a prescription for economic prosperity that would secure the future of Lihir well after the last mining manager had packed his bags. The Destiny Plan was a political manifesto and economic strategy *for Lihir,* effectively based on the total implementation of the PV course and its philosophies throughout Lihir and eventually the nation.

When Filer and Jackson first advised Kennecott on the potential risks to the project, they suggested that resentment (towards the State, authorities, business, Europeans, or simply economic inequality) prevalent within the Nimamar Association and earlier movements could dissipate under the impact of mining, but that these same feelings might also resurface and find expression later on in a more 'rational' (yet potentially violent) disguise (Filer and Jackson 1989: 177). Feelings of resentment throughout Lihir may not have surfaced in the type of violence witnessed on Bougainville, but they have certainly been expressed within a more 'rational' manoeuvre headed by the LJNC. Those aspects of micro-nationalism that stress some form of loosely defined 'ethnic cohesion' or community membership based on common goals were familiar traits of Nimamar. As we saw in the previous chapter, mining operations did amplify these sentiments, but they have not been channelled into direct political activity, instead finding expression in daily relations between Lihirians and non-Lihirians. However, where Nimamar tapped into a feeling of discontent, the LJNC tried to raise these feelings to fruition, from a micro-nationalist attitude, to an actual attempt at micro-nation making.

Finalising the Review

In mid-2004, the LJNC visited Kairiru Island near Wewak for an intensive retreat with Samuel Tam. During this time the LJNC members consolidated many of their ideas and signed the 'Kairiru Accord' in which they expressed their commitment to the Lihir Destiny. The LJNC returned and recommenced negotiations with renewed vigour. The Lihir Destiny would be a 20-year process that would fulfil the National Goals. The LJNC would lay the foundations, and then move to 'import replacement', 'industrialisation', and finally the 'information and technology and services era'. Yet it still struggled to get the support of the company and the State. While there was generally agreement over the actual compensation agreements, there was little accord over the expenditure of funds for PV, the new governance structures, or the neat models for social and economic development. The company refused to allocate funding for plans that appeared unstable and vague, as well as being poorly received by the wider community, while the State wavered between applauding the LJNC's plans for self-reliance and dismissing its members as 'cargo cultists'.

The company had made an unprecedented offer. They were committed to spending K100 million over five years until the following review. This included the entire range of compensation payments and development of infrastructure and service provision. The LJNC rejected this offer, claiming that it would take at least K4 billion to develop Lihirians and to 'balance' the disturbance from mining. With their new awareness of 'holistic human development' from PV, the LJNC wanted the new IBP to meet *all* the needs of *all* individuals.

Both of these figures were equally arbitrary. There was little basis to the company's offer other than its impressive ring. The LJNC's ambit claim wasn't just a copycat performance of Francis Ona's infamous demand for K10 billion for damages caused by the Panguna mine, but it was a bargaining device. The LJNC members realised that demand was unrealistic, but they wanted to express their commitment to PV and their Destiny Plan. They argued that, if the company could not meet their demand, then the only way forward was to support their vision:

> The answer for the huge differences is not the mentality of LMC dishing out cash to the community to satisfy their needs nor in the mentality of depending on LMC giving us K20 million per year to satisfy our needs.

> We believe that the answer to this problem is with the people themselves. They must make their own money to satisfy their own needs (Physical, Spiritual, Financial, & Emotional Needs). We believe this is the reality of life to follow God's Order to Adam and Eve (LJNC 2004b).

Recourse to the Fall, as justification for the hardships that people must endure along the road to prosperity, rationalises and converges the morality of capitalism and personal transformation as it is set out within PV. What we can also see from these contrasting offers and counter-demands is the extent to which both parties were approaching compensation from different angles. For the company, compensation is proportional to the level of destruction, and once payment is made, all on-going obligations are negated; for the LJNC, compensation is aimed at developing novel and on-going economic relationships (Jackson 1997: 106–8). The company argued it was already exceeding its obligations to provide for community development and service provision, while the LJNC appears to have interpreted the *Mining Act* to mean that the company should fund anything regardless of its scope or nature.

In effect, the LJNC members were stating that they would develop Lihir in isolation, and it was the company's responsibility to help facilitate their plans. In their words, 'the mining company was the stepping stone for the Lihir Destiny'. The company was concerned about what appeared like a politico-economic manifesto with serious implications for the community and the potential to bring the company into conflict with the government. But the LJNC members were resolute: they were convinced that they had finally discovered the key to economic prosperity and refused to adjust their plans.

In many ways, the LJNC members conflated management issues with their desire for ownership. Despite their endless production of complicated flow charts and economic projections, they had not come to grips with the problem of managing a multinational operation, or an 'independent island nation', assuming that operations would be unproblematic once they were in a position of control. Like so many of the issues which arise in the context of mining, such as compensation for environmental, social and political impacts, the Destiny Plan should be viewed as a local response to conflict over the control of resources (Banks 2002). Controlling their own lives, the mine, access to wealth and benefits, and the direction and type of economic development generated through this process, was central to their vision of the future.

The negotiations frequently required an independent third party arbitrator, and some senior company managers admitted that they simply wanted to complete the review and move on. Eventually, a settlement was reached and a new agreement was signed in 2007 (see Plate 6-2), partly because it was difficult for the company and the State not to support a local commitment to economic self-reliance — effectively an indigenous model of development. While there were doubts about the details and the means to reach the new destination, it was assumed by all parties that, once an agreement was made, planning and implementation would be worked out along the way.

Plate 6-2: Mark Soipang signing the Revised Integrated Benefits Package (the Lihir Sustainable Development Plan), Potzlaka, 2007.

Photograph courtesy of the LGL Archives.

Lihirian leaders made some headway in fulfilling earlier desires for minimal State involvement. Lihirians maintained their 50 per cent share of the 2 per cent mining royalty collected by the State, with SML landowners receiving 20 per cent in cash payments, the Nimamar Rural Local-Level Government and the Nimamar Special Purposes Authority receiving 20 per cent and 10 per cent respectively, while the other 50 per cent went to the New Ireland Provincial Government. The provincial government would receive 70 per cent of the Special Support Grant, while the remaining 30 per cent went to the local-level government. The LSDP builds on the format of the broad definition of compensation in the original IBP, with five chapters that address specific issues, projects and phases of mine life — The Lihir Destiny, Destruction, Development, Security, and Rehabilitation. The budget was later increased to K107 million to be portioned out between the various chapters according to priorities.

These chapters are set in context by a mission statement that outlines the time frame for achieving the Lihir Destiny over the next 20 years, beginning with holistic human development, economic development for communities, entrepreneurial class development, and self-reliance and financial independence. Underscoring this is the definition of the Lihir Dream and the spirit of the IBP which is based on the following objectives listed in the Executive Summary:

a) Parallel Development: To ensure that development in all villages in Lihir will happen in parallel to the development of the Lihir Gold Project.

b) Balanced Development: To ensure that development in Lihir is balanced in all villages and wards in Lihir.

c) Sustainable Development: To ensure that development in Lihir is sustainable. That is, that development in Lihir must be able to sustain itself without being dependent on the Lihir Gold Project.

d) Stable Development: To ensure that development in Lihir is stable. This must happen in harmony with the Lihir Society and not destroy and erode the order and culture that existed in the society prior to the operation of the Lihir Gold Project (LSDP 2007).

Chapter 1 is supposed to facilitate projects that build the foundation for achieving the Lihir Destiny of a 'healthy, wealthy and wise society' — such as health, education, and law and order programs. This includes institutional capacity building with the local-level government and its various agencies, revamping health and education services, and the further development of community infrastructure such as roads and reticulated water and electricity. The generation of local media outlets, such as newspapers and radio, is also given priority. Chapter 2 is concerned with support for the LMALA and those agreements specific to the loss of land for mining activities, and the continued delivery of benefits and compensation to those who own land in the SML area.[10] Chapter 3 outlines proposals for alternative economic development, such as livestock and agricultural projects, aimed at reducing reliance on an industrial economy. This chapter also contains details about contracts around the airport, the mining pit and stockpile, the plant site, and the (continuing) relocation of Kapit village. Chapter 4 is primarily devoted to 'Holistic Human Development' to be delivered through the PV course and the PV Home School Program, and Chapter 5 lays out initial plans for mine closure as required by national government legislation.

Miming the Mine

To fully understand these plans for achieving a desired state of modernity, we must appreciate the centrality of mimesis. The repetitive rote learning of PV laws, mantras, anthems, disciplines and entrepreneurial business techniques was aimed at guaranteeing access to a certain kind of life and status. For PV followers these 'rules' were commonly seen as ends in themselves. Following correct PV procedure often seemed more important than devising ways to achieve desired entrepreneurial goals that might actually draw upon useful skills learnt in the PV course. This partially accounts for the gap between theory and practice. For the LJNC, this was manifest in a Destiny Plan that was completely committed to the exact performance of Tam's teaching. Tam had become the unwitting or reluctant charismatic leader of a post-modern cult.

Tam's educational system, with its ranked hierarchies, reproduced the effects of graded societies that are intended to give people access to new forms of knowledge or 'secrets' as they are initiated into higher levels. The LJNC members spoke as if 'sitting down' in the highest level PV course (the Entrepreneurial Business Development course) would somehow reveal the answers. Undoubtedly, this was compounded by the fact that Tam literally said that he was letting them in on a 'secret' — his knowledge gained from years of business achievements that LJNC members had not experienced. When one of the LJNC members triumphantly exclaimed that they had 'unlocked the White man's secret', I momentarily tried to believe that it meant a deeper insight into the self-contradictory and exploitative ideology of modern bourgeois illusions that would actually locate me in the 'savage slot'. This was mere academic fancy; this LJNC member's newly discovered knowledge of tax evasion and investment strategies was probably of more immediate use.

---

10   Separate from this agreement is the funding of LMALA, which in 2009 received approximately K5.2 million from LGL for all operational expenses.

'Cracking the code' is a familiar aspect of cargo discourse (Worsley 1968: 247). The PV course appeared to provide the answers and knowledge seemingly withheld by the company and the State. Tam's entrepreneurial instructions focused on many of the basic managerial elements of business, such as conducting feasibility studies, drawing conceptual flow charts, and regularly producing budgets, report sheets and sales projections. These and other unstated and implicit aspects of business, which underpin the operational success of the mining company, were initially treated by the LJNC members as a form of hidden knowledge which they could make explicit through the routines of naming and designation, and through ritual enactment of the procedures of business and the tacit 'rules' that govern Western personhood.

In his work on millenarian movements in New Britain, Andrew Lattas argues that humans constitute themselves by internalising and reacting to the gaze of others. In cargo cults, this assimilation of the other is manifest in the 'theatrical self-identification in which people's corporeal gestures become identified with those of Europeans' (Lattas 1998: 268). To this end, the LJNC's plans for economic, social and political advancement represented the appropriation and internalisation of the 'moral technology of self improvement' (ibid.: 234) consistently presented by kiaps, missionaries and miners. Personal Viability and its application in the Destiny Plan were attempts to overcome the alienating structure of modern economic and political relations. Yet these strategies seem to simultaneously reproduce the very structures and organising principles they sought to reform (see Comaroff 1985).

The LJNC members were not mistaken in seeing the regimented bodily routines and gestures, disciplined business techniques, and quests for personal augmentation — all taught by PV — as the 'secret' or essence of economically successful modern Western lifestyles. Indeed, even cargo cults often consist of perfectly rational economic and political action (Dalton 2004: 203). Their emotional connection to PV and their Destiny Plan stemmed from their attempt to grasp the transformative powers of discipline. It should come as no surprise that the militaristic character of PV and the Destiny Plan reflect the same pedagogic regimes used by administrators, missionaries and teachers to create disciplined moral subjects — more civilised Melanesians (Foucault 1975). We can detect a certain element of 'rigorism' in the LJNC's plans for total imposition: the tendency among millenarianists who believe that they now hold the key or have grasped the secret and feel compelled to force it upon others (Burridge 1971: 135). For the LJNC, economic prosperity was dependent upon all-encompassing behavioural regulation.

The overwhelmingly bureaucratic and autocratic character of the Destiny Plan could not have developed without the daily influence of the company's regimented constitution and infrastructure. Mining companies invariably fit the

Weberian ideal type model of the bureaucratic enterprise, with their integrated management systems (Weber 1978: 956–7). It is this conflation of modern structures with economic success that caught the LJNC eye. As Lihirians have been progressively drawn into greater contact with the outside world, they have familiarised themselves with new things, movements and concepts which they incorporated into their way of operating. However, their knowledge of the outside world and global capitalism remained partial. In earlier years, Arau was the main external conduit: he was the chief source of information on life outside Lihir, political developments occurring throughout the province, and visions of a possible future. The mine has provided access to a range of external sources of information through media, telecommunications, and the constant movement of people on and off the island. Over time, Lihirians have filtered more information in a different sort of mimetic behaviour.

The TKA envisioned its economic future as lying in plantation agriculture that was no longer managed by their colonial *masta*. These plantations would be collectively controlled and worked by groups of Lihirians. This vision incorporated idealised notions of tradition and current ideas about cooperatives. At the time, cooperatives were seen as the only option for economic advancement that also allowed for the incorporation of an egalitarian ethos. In the context of mining, many Lihirians see their economic future in the management of the mine and in small-scale businesses that capitalise on mineral wealth. Lihirians are again seeking to displace those whom they see as dominating them — white mining managers and businessmen from other parts of PNG with lucrative contracts. The egalitarian ethos once again appeals to tradition, but the model no longer incorporates ideas about collectivism. It is appropriate that many Lihirians imagine themselves operating within a post-colonial economy as individual businessmen and managers.

Making Myth-Dreams Come True

Without ignoring or denying the Lihirian passion for moral equivalence — a fundamental aspect of their desire to manage their own lives and the mining project at the same time — we should not miscalculate the material dimension of the *Lihir a ninambal* (Lihir dream). People retrospectively interpreted this dream in different ways. Some recognised the mining project as the logical fulfilment of local prophesies, but for most Lihirians who were not in receipt of mining royalties, compensation payments or relocation houses, new inequalities testified to the falsehood of recent reckoning. Some assumed that the arrival of the company, and especially its early supply ships, was the realisation of predictions that ships and planes full of cargo would arrive from America, but it was the greedy and selfish actions of a few (such as the Putput landowners or the government) that spoiled this for everyone. This echoes Bah Arom's comments at the beginning of Chapter 3 about Kennecott's arrival and other nostalgic recollections of the Ladolam mining camp. This romantic period in the relationship between Lihirians and the company played a strong role in encouraging the belief that Lihirians 'called' the company to their shores, partly explaining the common conviction that they own mine.

Underscoring the Lihir Destiny charter myth is the term *a peketon*, which gained metaphorical significance in the TKA movement. It became the leitmotif of the Destiny Plan: national change will emanate from Lihir, providing moral justification, connecting modern means with traditional dreams of influence and change. On the one hand, the LJNC rejected all forms of thinking which supported these earlier prophesies or myth-dreams, claiming that these hold Lihirians back from 'real development', fostering a handout mentality. On the other hand, the LJNC gave credence to these dreams, intellectualising them as Burridge would, recasting them as aspirations, and through the PV lens as the Lihir Destiny. The LJNC members seized upon prophesy fulfilment to muster political support throughout Lihir, historically contextualising their plan to the extent that they too came to believe that becoming a city (the outward manifestation of independence and self-reliance) was truly their destiny.

Plate 6-3: Signboard outside the Personal Viability Training Centre where Sam Tam resides in Marahun townsite, 2010.

Photograph by the author.

Myth-dreams in the TKA promised that ancestors would return bearing cargo and money and give houses to the faithful. Some of these beliefs have continued to circulate throughout Lihir, finding greater currency among the disenfranchised. Burridge's solution for these myth-dreams or esoteric aspirations is a staunch commitment to modernity, to be delivered by the better behaved European, or in the Lihir case, the moral mining manager. But the LJNC members were simultaneously modern myth-dreamers and enlightened moral Europeans. They had (re)created their own myth-dreams in the form of the Lihir Destiny, while simultaneously rejecting the original myth-dreams as they became the moral agent of change. While the colonial administration may not have been successful in its endeavours, the LJNC proudly bore the White Man's Burden, 'out of love for the people', in its effort to save Lihir from financial — and moral, customary, spiritual, emotional, physical and mental — destitution. As one LJNC member proudly explained to me: 'I will act like Jesus did, he died on the cross, to redeem everybody, mi, mi laik halvim yupela olgeta [I want to help all of you]'.

The Road Ahead

If the original IBP was once heralded as a new industry benchmark, then the Destiny Plan (the LSDP agreement) could well be the new yardstick for such agreements — if not for the scale of the package, then for its ideological and practical complexity that is so perplexing to company managers, the State and Lihirians alike. After the signing of the agreement, the LJNC was dissolved and a new LSDP Planning Monitoring and Implementation Committee was formed under Soipang's leadership. This was supposed to be a roundtable affair with representatives from the LMALA, the local-level government, the women's association, the Church and LGL. In theory, this governance arrangement represented a shift towards greater local participation and responsibility, with less emphasis on corporate service delivery. Money committed under the LSDP is to be used on a range of projects, guided by the vision of economic independence for the entire Lihirian community.

In reality, implementation has been consistently stymied by institutional incapacity within the committee itself and legal battles over the LSDP agreement between the local-level government and Soipang, who has continued to leverage authority vis-a-vis his role in the LMALA. Part of this struggle stems from ideological differences concerning how future visions will be achieved, surrounded by questions over who actually holds the legal mandate to assume leadership and representation — the democratically elected leader or the self-appointed visionary? Not all of the governance structures outlined in the early conceptual plans — such as the Lihir Grassruts Pawa Mekim Kamap or the Kastom Lidas Komiti — have been put in place, and for the time being the codification of *kastom* has been abandoned. However, the emphasis on PV has remained and the new committee has continued to institutionalise it. The plans within the LSDP have evolved, but there has been a determined concentration on business development, town planning, and health and education services.

The social risks and hazards systematically produced as a result of mining activities, which consequently threaten operations, obviously need to be minimised and prevented, or at least channelled. The LSDP is a progressive response, but it has also generated a complicated and confusing corporate social responsibility legacy. Aside from the seemingly impressive financial commitments, the LSDP is best characterised as a work in progress, or an evolving strategy for sustainable economic development. In its current manifestation, LGL and Lihirian leaders are no longer just dealing with traditional questions over the legitimately unequal distribution of wealth, but with the challenge of implementing and sustaining techno-economic development. To be sure, the LSDP has not defused any of those ticking time bombs once predicted by Sir Julius Chan; it may have actually generated an entirely new set of grievances that are yet to surface. But in the corporate environment, it is a sufficiently thick anti-flack jacket to guard against claims of non-compliance with international standards of corporate social responsibility. Of course, if the LSDP proves unworkable and Lihirian frustrations boil over, then the current version of the LSDP will be a pretty flimsy defence of mining operations, and the government may well find itself faced with a more sophisticated version of past events on Bougainville. While the company and the new LSDP Committee have struggled to implement the LSDP agreement, my concern has been with how this agreement came to represent a very specific cultural response to mining which captures the changes, tensions and contradictions in Lihirian engagements with modernity and local people's desire for social and economic advancement.

7. Custom Reconfigured

> The claim is not that culture determines history, only that it organises it (Sahlins 2004: 11).

> *Kastom i no inap dai, em bai stap oltaim* [*kastom* cannot die, it will be here forever] (Paulus Tala, Puki hamlet, Lihir 2004).

Throughout my first weeks of fieldwork in late 2003, I was immediately drawn into a web of mortuary rituals and *kastom* politics. Several days before I had landed on the island, a young man named Stanis Kanpetbiah from Lesel village had unexpectedly died. When I eventually found my way to Lesel, I made my way into the men's house and sat with the men from Tiakwan clan who had adopted me during my previous visit. This was home for several weeks: here we ate, slept, socialised and remembered Kanpetbiah. Throughout this mourning period, pigs, shell money, cash, garden produce and trade store food were exchanged and consumed. Women prepared food for guests while senior men sat inside the men's house talking, smoking and chewing betelnut. People were expected to follow mortuary taboos that maintained the sombreness of the occasion. Respect was being paid to the recently departed, through processes that 'finish the dead' and continue the lineage through a gathered sociality. My introduction to Lihir gave me a glimpse of an institutionalised form of social reproduction embedded within the men's house. This was, in the words of one man, *pasin bilong Lihir* (the ways of Lihir).

This occasion also revealed new processes, values, practices and tensions. Not everyone from the lineage stayed for the entire duration. Males travelled around, and went to work in the mine, or came to socialise for a few hours at night before retreating to their own house to sleep, while some men drank on the sly, now that beer was considered an integral part of any customary gathering. Women negotiated keen divisions between kinship lines made apparent by the unequal distribution of wealth between the assembled clans. Senior males conspicuously harangued younger males as they addressed new disputes and deviations from men's house ethos, concerns over the correct ways to perform *kastom*, and the need for clan unity. A Catholic communion service was held in the men's house, followed by the distribution of Kanpetbiah's possessions. There was a level of ambivalence and dispute over the dispersal of his savings and whether or not it was right for people to keep a record of the store-bought contributions to the feast — whether gifts bought with money should be forgotten or reciprocated. After several weeks, the mourning period was finished, the taboos were lifted, and the remaining males launched into a spectacular drinking binge, spending hundreds of kina on beer and spirits. They drank well into the following day, stating that this was a time for *hamamas* (happiness).

At the time, I wondered what was implied by *pasin bilong Lihir*. The trade store food, the blue plastic tarpaulin shelters, the trips to the market in hired vehicles to buy more tobacco and betelnut for guests, the ferrying of recently purchased pigs on the back of trucks, each one's presence announced with a blast from a car horn, and the excessive drinking, made the event appear like a parody — not quite modern, but certainly not traditional. However, this was a socially significant event. Meaningful transactions were taking place, and relationships and collectively held values were expressed and maintained. Introduced goods and styles did not detract from the purpose of the occasion. Indeed, they helped to elicit sociality on a grander scale.

Beyond the absence of thatched huts, traditional goods and ornaments of rank, there were perhaps some ways in which this event still reflected the façade of contemporary *kastom*, if not a deeper reformation of society. This occasion was acceptable to participants and served intended social purposes. But it did so through processes in which contradictions of meaning were unintentionally and unconsciously concealed or veiled. This was not 'spurious tradition' (Handler and Linnekin 1984) or inauthentic culture, for to be sure, tradition is not stasis but a particular way of changing in new circumstances. But the performance of Lihirian *kastom* presents a seamless continuity with the past that masks disjuncture and fears of declining sociality, which raises important questions about how modern mortuary feasts retain their status as the embodiment of tradition and *kastom*. In this chapter, I look at the ways in which Lihirians engage with *kastom* in both conversation and in practice, and how this is related to their own responses to extreme social and economic changes and the ideas, visions and plans of the new political elite — the very particular modernity promulgated in the Lihir Destiny Plan.

Understanding *Kastom* as Culture and History

When Lihirians use the term *kastom*, it can assume a variety of meanings depending on the context and the intentions of the speaker. It can refer to those activities, beliefs, values and forms of behaviour that are also glossed as *pasin bilong tumbuna* (the ways of the ancestors), but it in no way encompasses everything their dead relatives once practiced. In recent years, the Tok Pisin word *kalsa* (culture) has entered Lihirian discourse. It derives from the English word culture, but like *kastom*, it often refers to certain values and practices, usually associated with an idealised past, rather than culture *per se* — all the cultural intimacies of daily life that anthropologists are so fond of documenting. Lihirians alternate between these terms, but generally they can both be seen as descriptive and conceptual tools for referring to the past and assessing the present. At the national level, the potency of these terms derives from their vagueness (Narokobi 1980; Keesing 1982: 299), but in Lihir, it is their specific meaning that gives them such rhetorical strength.

If we are to maintain that culture is always in a state of flux and never static, and that change can originate from both inside and outside a community, then we must recognise the efficacy of 'inventions' and interpretations of culture which occur in the present (Wagner 1975). Thus *kastom* is the product of interpretations of the past by present generations as much as it is the residual essence of the past which has trickled down to the present. In Lihir, *kastom* is a feature of the discourse of change. At times, its referential meaning can appear vague and encompassing, but its reflexivity allows for comparison and contrast with other sets of cultural practices. *Kastom* cannot be viewed as an unconscious cultural inheritance, as if it is somehow distinct from a self-conscious proclamation of the past in the present. It is a self-conscious construction of the past which is used to inform present behaviour and a sense of identity.

The emergence of the discourse of *kastom* is not solely the result of recent change and development; it has been present since at least the late colonial period. However, it has gained a greater *ideological* stronghold since mining began. This is part of an ongoing process of cultural reflexivity characteristic of the experience of modernity. It also reflects the more general Melanesian phenomenon wherein specific ritual practices and forms of sociality are reified and held in contradistinction to local images of Western society, politics, economic practices and personhood (Bashkow 2006).

While anthropologists have commonly approached *kastom* as a rhetorical creation at regional, national and local levels, we should do well to consider Akin's argument that anthropologists might have reified it more than Melanesians (Akin 2005: 185). When *kastom* is severed from its cultural moorings, we are likely to overlook the intense interaction between culture and *kastom* as both continually shape each other over time, and thus be led to neglect 'the concurrent subjectivisation of *kastom* as culture' (ibid.: 186). This is important to remember because it reinforces the point that *kastom* is not merely the essentialisation of Lihirian culture, but part of Lihirian people's 'culturally specific modes of change' (Sahlins 1992: 22).

Mining has exposed Lihirians to alternative forms of living and relating that have challenged and destabilised existing ways of life over a very short period of time. As a result, many Lihirians look to *kastom* for social stability, which has generated a sort of mass 'custom cult' whereby Lihirians spend incredible amounts of time and resources ritually reaffirming their belief in *kastom* as the true road to a harmonious modern existence — the enactment of the develop*man* project. Notions of *pasin bilong Lihir* consciously reinforce this belief and form part of the ideology of *kastom*, which in the Marxist sense distorts or conceals particular realities. Social relations have been reified to the extent that *kastom* is held in opposition to the road to development outlined in the Destiny Plan, which is based upon the peculiar ideology of Personal Viability, or *bisnis*. But as we have seen, this is further complicated by the fact that the Destiny Plan opposes and simultaneously incorporates both the ideology of *kastom* (by rejecting the develop*man* process and proposing a codified set of *kastom* rules) and the ideology of landownership (by rejecting dependency upon compensation whilst demanding corporate funding for an alternative grand scheme). Moreover, the ideology of landownership requires the specification of land rules to determine who can receive compensation, which in Lihir is inseparable from the codification and the practice of *kastom*. We can thus begin to see how *kastom* directly emerges from the legal bowels of resource development, and the tensions and ambivalences in the different ways that Lihirians conceive of development and *kastom*.

The ideology of Lihirian *kastom* has been highly permeated by a Manichean allegory comprised of contrasting opposites which help to emphasise the specifics of each category. Missionary discourse has been internalised throughout much of Melanesia, as people come to see their world and their history as a set of oppositional contrasts that depict struggles of light and dark, goodness and evil, God and Satan, indigenous and exogenous, tradition and modernity, or *kastom* and *bisnis* (Kahn 1983; Errington and Gewertz 1995; Foster 1995a; Robbins 2004). In some cases, development is even conceptualised as 'rebirth' (Keesing 1989: 27). Paradoxically, in many instances, these inherited postulates have been inverted in a new Manicheanism: instead of associating *kastom* with the heathen, people are expected to preserve their ancestral values and to keep the temptations and 'cultural sins' of Western life at bay. Such dualisms assist the objectification of traditions and customs as something 'thing-like', to be (literally) separated from people and communities, reduced to written form, or made into a budget line item, an 'LJNC portfolio', or a new category to be ticked off on the Personal Viability checklist along the road to the Lihir Destiny.

The doctrine of *kastom*, especially as it is manifest in Society Reform and the Destiny Plan, requires the exclusion of 'non-indigenous' things. However, the difficulty of enacting this doctrine is mirrored in the struggle to retain local business as a separate sphere of activity. Thus the inability to implement reified Western relational and economic practices in daily contexts is deeply entwined with the complexities associated with segregating *kastom* from capitalism, whether it is *bisnis* or the corporate mining economy. In practice, the conflation of these supposedly separate spheres constantly compromises the pursuit of both. When taken too literally, these cultural polarities ignore crucial interpenetrations and fail to capture the actualities of the Lihirian economy. As modern consumer goods are poured into Lihirian exchange, and are often preferred over those of traditional origin, *kastom* generates an increasing demand for them, reinforcing dependency upon capitalism more generally. Ironically, we then find that capitalism actually 'energises' *kastom* rather than subtracting from it (see Thomas 1991: 197).

Staging *Kastom*

In July 2004, I became involved in preparations for the mortuary feasts being performed by members of the Lamatlik clan who belonged to the men's house in Natingsangar hamlet in Kinami village, into which I had become incorporated. This feast, which was one of the few occasions on which I was more involved as a host (*hurkarat*) than as a guest (*wasier*), was the next stage in fulfilling obligations to a recently deceased clansman, Umbi, and two other men, Piong and Rapis, who were still alive. This *katkatop* was led by Bah Arom, his maternal nephews John Zipzip, Clement Papte and Peter Toelinkanut, and several of their sons and male affinal relations who resided in the same hamlet.[1]

At first glance, this is an unlikely event on which to concentrate. It appears to refute ethnographic orthodoxy: it was neither ritually nor politically climactic. The 'deceased' were not yet 'finished' — indeed, two were still alive. However, through comparison with a later *katkatop* (or *pkepke*) feast in Putput in July 2006, we can gain insight into common *kastom* practices, including the ways that men stake their claims, make rules, and reproduce their standing in society, and into the collapsing or merging of feasting stages. Through comparison, we can gauge the excessive nature of the Putput feast, which demonstrates the way that differential access to wealth influences the performance of *kastom* and also the difficulties with codified ideals for both landowners and non-landowners. My intention is to convey a sense of the performance of *kastom* as temporally contextualised social action, rather than to provide a processual account or an ahistorical tabular summary that privileges structure over practice. In short, these two feasts highlight the point that, while people happily offer authoritative versions of the sequential structure of mortuary feasts, every

---

1   The men's house that I was affiliated with through my incorporation into Tiakwan clan is located in Lesel village, the site of the first mortuary feast I attended. However, I was more closely aligned with the men's house in Natingsangar hamlet in Kinami village, where I resided. My main patron, Francis Bek, is the son of Bah Arom, the owner of this men's house. In Lihir, it is common for at least one son to remain in his father's hamlet. As a result of my connection to Bek, I was closely involved with Arom's nephews and the activities that took place in their men's house.

performance reflects an improvisation of general procedures. Mortuary rites might be sequential, but the outcomes are not axiomatic or unproblematic. Various factors will always determine the results and the reception, ensuring that mortuary rites remain a contingent and flexible practice.

Moreover, it is common to see *katkatop* rather than final *karat* feasts being performed, partly due to the relative difference in scale and ease of mobilisation. There are two types of *katkatop*, both essentially the same feast with the same purpose, the difference being whether the celebrated person is still alive (*katkatoptoh*) or dead (*katkatopmiat*). The first feast in Kinami was a combination of the two. Both types are directly concerned with the inheritance of political power and resources within the lineage and clan. *Katkatoptoh* is common, partly because there is more at stake. If the 'deceased' (who is still termed *kanut*) witnesses the host's efforts, then there is a better chance that the host will be rewarded. The same applies to final *karat* feasts. The cumulative result of these changes is the collapse of sequential feasting time. People have to make *kastom* sooner, while relatives are still alive, to reciprocate existing debts from other 'prematurely' performed feasts, to secure their own resources and customary influence, and to maintain prestige within this sphere. In effect, the availability and integration of cash has enabled competition to be 'brought forward', and has transformed what was formerly a post-mortem competition between 'heirs' into a regular ritual performance of the political stratagems that permeate contemporary village and lineage power struggles. While mortuary ritual has always been partly driven by economic and political factors and the need to fulfil obligations, there is a structural and symbolic transition as greater emphasis is placed on money, resources and competition, displacing the centrality of the 'deceased'.

Making *Katkatop* in Kinami

Zipzip and the others had been preparing for this event since the death of their clansman in 2003, but within weeks of the planned date there was still confusion as to whether things would go ahead as scheduled. Not all of the necessary pigs were secured and much of the food would still have to be bought. Extra gardens had been planted in preparation, but they would not provide enough yams to cater for the entire event. Some talked about postponing things, while others insisted on proceeding. As the opening day drew closer, it was becoming obvious that we would need to travel abroad for our needs. Three days before the intended start, two dinghies were sent to Tabar to find suitable pigs. Zipzip and Toelinkanut often insisted to me that locally raised pigs should be used because they carry more prestige, but the inflated demand could not be met internally.

The feast was to be held over six days, and would combine parts of the *rahrum* feast that had not yet been made for Umbi and Rapis. This combination was largely for reasons of prestige, practicality and necessity, and to accommodate the constraints of employment. Clement Papte worked in a relatively senior position in the mining company's Human Resources section and had taken leave for this feast. With so many last minute preparations, and the lingering doubt over the number of pigs, he was anxious about whether the event would begin before he had to return to work. Several others were bound by similar commitments, while some could only attend during the evening.

Another three boats were used in addition to the first two, and by the Sunday evening before the event was due to be begin, all five dinghies arrived back with a total of 19 pigs. All of these pigs were transported from Londolovit wharf to Kinami on a large flat-tray truck that had been hired at great expense from a neighbouring village. Monday morning was spent putting these pigs into temporary caged enclosures (*garum*), collecting the final pieces of firewood, and husking green coconuts for the men's house. This work was performed collectively and it was expected that other groups within the village would assist as a sign of inter-clan solidarity. Women cleaned around the men's house, tidied the grave sites within the men's house and put flowers on Umbi's cross. Rice and tinned fish were cooked for workers (hosts) and guests who arrived early. Another three boats were hired and sent to Masahet to collect other relatives and garden food. Preparations were made for the *kienkien* feast which acknowledged the preparatory work and opened the main feast.

By Tuesday morning guests were still arriving. Over the course of the week, some 300 people, including women and children, would attend. Not all of them would stay in Natingsangar, as many were either from Kinami or travelled from the surrounding villages. In Lihir, guests are not directly invited; instead, people come to demonstrate support and alliance. Once the hosts have announced their intention to hold an event, news spreads quickly, and it is expected that people in the neighbouring villages will attend, reflecting the open nature of feasting and Lihirian attitudes towards hospitality.

The advent of motor transport has significantly changed the role of the hosts, which used to require the construction of large temporary houses for visitors, both inside and outside the men's house enclosure. Increased availability of transport means that guests can simply attend on those days (or in many cases for those hours) when food is distributed and consumed. Previously, people stayed with the hosts for several weeks or even months, increasing the debts and obligations between guests and hosts, giving each party greater purchase on the other's resources. For all the talk of declining hospitality, not everyone has been disappointed by the changes. The drain on resources and the uncertainty about when guests will leave can prove burdensome. Although the duration of the feast has shrunk, the nature of contemporary *kastom* ensures greater expenditure over a shorter time.

On Tuesday afternoon, a single pig was killed and cooked in the men's house for the *kienkien*, while women cooked vegetables outside. Zipzip made a speech to announce the activities for the following day and to acknowledge the beginning of the feast. Guests received portions of cooked pork and yams with their *pinari* (gifts of betelnut, pepper sticks, green coconuts and tobacco), which had mainly been purchased at the market that morning.[2] By the evening, the shelters were brimming with women and children, while men and young males slept in the men's house and other shelters constructed for the occasion. The hamlet was abuzz with chatter, screaming children, chastising mothers and packs of dogs seeking scraps.

On Wednesday, the *rarhum* commenced. Not all of the stages were performed, only the most important — the *balunkale* and *berpelkan*. Visually, the feast was distinguished by clan heirloom *mis* strung up in the men's house. There was some disagreement over the stages as some people later insisted to me that instead of *ber pelkan*, it was *balunpeketal* that was performed. The latter name refers to the pigs that are normally killed when the deceased is seated and decorated in the men's house during the first mourning period before burial. This was a minor point of discrepancy, but it demonstrates how feasting rarely follows an ideal format. Donors are not always aware of how their pig will be used or whom it will eventually commemorate. Guests are even less likely to be aware of all the stages or last minute changes. Consequently, feasts are 'judged' by different standards of 'correct procedure'. Ultimately all three pigs were considered part of the *rarhum* and were treated as *bualtom*, which meant that they were confined to the men's house.

On the Wednesday morning, guests were served rice, tinned fish and vegetables. Males from the host group began preparing the pigs, cutting the two pigs marked for *balunkale* into nine portions to be cooked and distributed to the nine matrilineal groups that attended. A third pig (the *berpelkan*) was specifically for in-laws and cross-cousins of the 'deceased' person who was being honoured. Women prepared vegetables outside to be cooked with the pork inside the men's house. Women made their own separate earth oven outside the men's house that would only be used to cook vegetables. When the pigs were uncovered, the portions of pork and vegetables were placed in separate piles on the leaf beds together with betelnut, tobacco and green coconuts to form what is known as

---

2   One of the hosts described this transaction as *zukulwasier*. Although the name seems apt, since it means 'food given to guests', it normally refers to a stage within a final *karat* feast when food is given out to please guests while they wait for the dances to commence.

pinarilam (the big gift). This is the usual fashion for *balunkale*, which is strictly *bualtom*. During the day, women were busy weaving baskets (*piar*) to be used in later food distributions, and did not uncover their food until later that night. The three boats that had been sent to Masahet to collect other relatives came back that evening, and another small truckload of people arrived from Kunaie village.

By Thursday morning, the pork had all been eaten. Ordinarily, *rarhum* feasts are concluded with *susulkwil* (washing the skin) — the customary baths taken after men have abstained from washing during the feast. This is typically a climactic moment that ends a very sombre feast.[3] In this instance, when the *bebeh* pigs were announced (pigs cooked in preparation for consumption on the final day), this signalled the end of the *rarhum* and the beginning of the *katkatop*. On Friday, the climactic day of the feast (*banien*), the *bebeh* pigs were uncovered and distributed for consumption. Four pigs had been killed: one (the *berpelkan* pig) was consumed separately by cross-cousins and in-laws of the 'deceased', while the other three, known as *iolnizenis* (the pigs that people come to see being cut), were consumed by the assembled guests.

Buying Sociality

Completion of this feast required that the remaining pigs be butchered and distributed for guests to carry back to their men's houses. Before this could happen, the pigs had to be lined up in front of the men's house (a process known as *pasuki*) and publically 'purchased', which is crucial to the succession of leadership and the inheritance of resources (see Plate 7-1). As one of the leaders of the feast, Zipzip walked to the centre of the hamlet with a list of names corresponding to different pigs. Moving from pig to pig, he called out the name of the donor and whom it was intended to honour. It can be argued that this is the most essential element of *katkatop* and *karat* feasts. Some men even suggested that, if nothing else within the feast succeeds, the event is still considered to be 'correct' so long as this process is complete, and the hosts then have a legitimate claim on leadership succession and control over clan resources.

---

3   In the past these feasts were distinguished by strict taboos that applied to all males. Aside from remaining quiet and showing respect, this also included prohibitions on washing. This relates to the significance of the pigs (*bualtom*), which should not be carried outside of the men's house at any time during the feast. Men were expected to refrain from washing off the grease from these pigs during the feast. If they had to leave the enclosure for any reason, they were not to wash. The *susulkwil* signifies the end of these taboos as well as the feast itself. The prohibitions on washing have been lifted, mainly because of mission and government influence in areas of health and hygiene. In some instances, those men most closely associated with the celebrated person in the *rarhum* still adhere to this 'rule'. Usually, there is a pig to mark this important stage. The area where the men wash (which includes the section of reef) is now under taboo (*mok*), the reef cannot be used for fishing, and women are prohibited from passing through this area.

Plate 7-1: Peter Toelinkanut and John Zipzip arranging the pigs in front of the men's house, Kinami village, 2004.

Photograph by the author.

In those instances where the 'deceased' is an influential leader and maintains control over clan land (a title known as a *tamboh wan a pour*), it is in this context that the mantle of control can pass from the 'deceased' to the host. While not all 'deceased' clan members are influential or have significant resources to be inherited, and not all hosts will succeed to leadership or inherit land, this is still an important opportunity to demonstrate leadership. Just as hosts usually organise themselves under a central leader, guests also align themselves under a leading man. The presentation of pigs is above all a statement about leadership, people's access to resources, and their ability to coordinate themselves under their big-man. Hosts do not provide all of the pigs presented on this occasion. It is rather their ability to manage relationships that is manifest in the number of pigs they can arrange for their allies or guests to provide, which either reciprocate existing debts or create new cycles. This is an equally important opportunity for guests to express their respect for the 'deceased' and their capacity as a group.

The public 'purchasing' of pigs is usually termed *ravo matanabual*, which refers to the display of the 'price' or 'eye' (*matan*) of the pig (*bual*), with fathoms of *mis* linked together and rolled up in a leaf wrapping (*ravo*). The recipient pulls the shell money out of the wrapping whilst moving backwards and uncoils it for public inspection. Payment can occur in either the men's house enclosure or the central hamlet ground (*malal*), and in some exchanges the 'buyers' simply walk over to the 'sellers' and toss the required amount of shell money at their feet or into their baskets. The most important thing is for exchanges to be publicly verified.

On this occasion, all payments involved a combination of cash and *mis*. As the transactions got underway, the crowd was drawn in, eager to see just how much had been spent and whether the 'prices' matched perceived values. Only pigs purchased in Lihir were exchanged in public. Pigs raised and provided by the host group or their allies incur no payment; guests who receive portions of pork are expected to reciprocate when they later host the same feast. Pigs contributed by hosts or their allies (regarded as *wasier*), which have been purchased from an existing exchange partner or someone else, must be paid for on the final day of the feast. When pigs are purchased locally, transactors make arrangements prior to the feast and, depending on when the pigs are used, payment then occurs after they have been consumed or before they are slaughtered (see Powdermaker 1971: 201). The problem with pigs purchased offshore is that they require outright payment, which means that the owner is generally not required to attend the feast and there is no public verification.[4]

The obligations created through these transactions exist independently of the indebtedness that is created (or repaid) when the pig is later put to use. The emerging image is one of continual management. Big-men must be able to manage their relationships with the people from whom they acquire pigs and the people with whom they exchange live pigs or the guests who receive cooked portions of pork. Thus, in addition to the debts that are reconciled or created between hosts and guests, there is a complex matrix of exchange relations surrounding an event. New and pre-existing relationships are continually being negotiated. The ability to purchase pigs with cash, without any ongoing obligation, reduces the extent of this network.

There were three types of pigs being butchered at this final moment: *katmatanarihri*, *karemiel*, and *puatpes*. The pigs that had been lined up in front of the men's house (*katmatanarihri*) were to be butchered and then distributed to the assembled clan groups to carry back to their men's house for consumption. These gifts must be accompanied by a woven basket of vegetables (*piar*), and it is assumed that they will be exactly reciprocated at later dates when these

---

4   In some feasts, donors who have purchased their pigs from elsewhere have made a public announcement about the price of their pig and where it was bought. In these instances, the announcement is often intended to demonstrate their purchasing power.

men's houses host the same type of feast. Such gifts represent the strength of the host clan and the unity of the men's house. Portions of *karemiel* and *puatpes* are respectively allocated to men and women who helped to organise the feast. *Puatpes* and *karemiel* are not necessarily lined up in front of the men's house. If there are not enough pigs left over for these purposes, portions of pork can be drawn from *katmatanarhiri*, but only after they have been lined and paid for. *Puatpes* are often contributed by guests, and hosts will generally not know how many are going to be 'donated'. Allies will use this opportunity to make unexpected contributions to put hosts in debt, or to repay existing debts to them, and to test the leadership of the hosts. These pigs, which must be reciprocated at later feasts, are the objectification of solidarity (or *berturan*), and they represent the level of respect that a host can command. In this instance, 12 *katmatanarihri* pigs were lined up in front of the men's house, including one *tinanakarat* (the backbone of the feast), and approximately 25 more were given as *puatpes* and *karemiel* by a range of supporters.

In many ways, this feast was considered to be successful. There was an abundance of food, the guests were satisfied, the weather was clear, internal politics were kept at bay long enough to stop the event from splitting apart, and the hosts had demonstrated their collective capacity.[5] Relationships between the hosts and guests benefited from this experience, and respect (*sio*) was shown towards the 'deceased' (*kanut*) and the assembled guests (*wasier*). Peter Toelinkanut was reluctant to admit that success was contingent on the use of trade store food, buying pigs offshore, and purchasing yams and other gift items at the local market. He conceded that this was necessary because of the short planning horizon, but the event certainly failed to match any codified ideals. If the feast had been unsuccessful, then this would not only have been an embarrassment, but would also have confirmed the absolute necessity for a return to 'true *kastom*'. For the hosts, a range of relationships and expectations were ultimately at stake. But obligations to the 'deceased' were effectively fulfilled, important social values were still in place, sufficient money was spent without too much ostentation,[6] and the event *appeared* to maintain ideas about protocol and tradition. Most importantly, pigs were lined up in front of the men's house to validate the actions of the hosts, while individual transactions reinforced notions of virtuous sociality. Although purists noted the deviations

---

5   Magicians are commonly engaged prior to feasts to ensure good conditions, an abundance of food, and the satisfaction of the guests. Competing or hostile groups and individuals may attempt to counter these efforts through negative sorcery. Problems with feasting are often understood as the result of sorcery attacks, and despite their commitment to Christianity, many people continue to understand certain events as being specifically related to sorcery, indicating a continuum of spiritual beliefs rather than a set of absolute categories.
6   Several of the hosts reckoned that K18 000 or more was spent on this event by the host lineage and its supporters. Most of these expenses were for boat hire and the purchase of pigs. Other expenses included truck hire, trade store food, generator fuel, and market items like yams, sweet potato, betelnut and tobacco for *pinari*.

from past practice, people also reminded me that *kastom* survives precisely because they were still performing such mortuary feasts. While the ideology of *kastom* might distort the representation of actual social and economic processes, it is the continuity of *kastom* that is uppermost in their minds. However, as we shall see in the following case, it is possible for *kastom* events to be successful in some ways and to fail in others.

The Essential Develop*man*: Pkepke Putput Style

During a later visit to Lihir in 2006, another member of the LJNC, who was also a wealthy landowner from the Likianba sub-clan of the Tinetalgo clan in Putput, was hosting a large *katkatop* feast, which is known as *pkepke* in this part of Lihir. During the preceding week, utilities from around Lihir, loaded with drunken males, regularly made their way to Putput to deliver pigs for exchange and consumption. Several of these pigs died from exposure before the event had even begun. Some people speculated that the donors were already competing over their contributions, and that the disregard shown for the pigs was simply a reflection of their wealth. On the final and most important day of the feast, more than 30 large pigs were presented by allied clans to honour the 'deceased', several of whom were still alive. As the pigs were presented, many of the male guests became increasingly inebriated. Various people reported that over six pallets of beer had been sold from beer outlets in Putput and surrounding villages during the previous two weeks.[7] Pigs were typically presented by very boisterous and inebriated groups of men, usually with gifts of garden produce, bales of rice, boxes of tinned meat and cartons of beer. They were met by an equally drunk and excited host group to perform the customary 'sham fight' (Plate 7-2) with *mis* that accompanies the presentation of pigs (Plate 7-4) and dances. As the two groups approached each other, individuals typically planted one foot in front of the other as they held up a strand of *mis* like a spear and rocked back and forth in a mock challenge, shouting out appropriate relationship terms such as *a berturan* (friends), *a berpelkan* (cross-cousins), *a bertman* (fathers and children), or simply *a ginas* (happiness), together with statements about whether their pig was repaying or creating a new debt. As the afternoon wore on these encounters became increasingly farcical as drunken men tripped over themselves in their stupor and grew more aggressive in their greetings.

---

7   There are approximately 120 cartons per pallet. In 2006, one carton of beer (24 bottles) cost approximately K140 at local beer outlets in Lihir.

Plate 7-2: Customary sham fight to greet guests, Putput village, 2006.

Photograph by the author.

Plate 7-3: Paul Awam exchanging shell money, Putput village, 2006.

Photograph by the author.

Plate 7-4: Guests blowing the conch shell to signal their arrival with pigs for presentation, Putput village, 2006. Note the dollar sign on the conch shell, another instance of the incorporation of cash in ceremonial life.

Photograph by the author.

The assembled pigs were lined up to signify each of the 'deceased', followed by exchanges of *mis* and cash (Plate 7-3). As the pigs were butchered, the place was transformed into a sort of apocalyptic haze with the excited actions of the hosts, the squealing of pigs as they were about to be suffocated, and the smoke and smells from the fires that scorched the pigs' bristles before they were cut. One of the hosts, Paul Awam, announced that, even though this was a *katkatop*, supporting groups wanted to contribute dances to the occasion. Lively dance troupes from Malie, Masahet, Matakues, Putput, and from a far afield as West New Britain Province, performed throughout the afternoon and entertained the masses (Plate 7-5). When a minor drunken fight broke out over sorcery accusations, the day effectively wound down. Nevertheless, with over 500 guests, and an estimated combined expenditure of K60 000, this group had facilitated an exceptionally large *pkepke*. Indeed, in some ways the event was considered to be too successful, which raised plenty of criticisms. As we rattled around in the back of a truck along the bumpy road home to Kinami, my friends keenly observed the deviations from *pasin bilong Lihir*, and could not help but comment on the demise of ritual feasting.

Plate 7-5: Dance troupe, Putput village, 2006.

Photograph by the author.

This event reflects the mixture of concerns over *kastom* and the waning significance of the men's house and its associated ethos. It captures the ways that landowners have succeeded and failed at the develop*man* project. The hosts would doubtless claim that this was a valid event: exchanges took place, money was spent, people came together, pigs were lined up in front of the men's house to honour the 'deceased', and were later consumed. But some reactions suggested otherwise. People were concerned that the men's house was no longer the sacred ground where big-men nurture younger males. Some were angered that excessive drinking is now part of *kastom*. Some even rejected the feast because it was too excessive. The high number of pigs purchased offshore, the presentation of trade store goods, the drunken celebrations, and the inclusion of dancing were all heavily criticised. Traditionally, dancing only occurs in *karat* feasts, and only the mournful *rangen* songs are sung during the *pkepke* feast. The energised festivities detracted from what should have been the sobriety of the event, hence blurring the emotional distinction between *pkepke* and *karat* feasts. But the most damning critique came from people who boycotted the event because they knew that it was funded by mining royalties. It represented the arbitrary distribution of wealth rather than the actual strength of the clan or the hosts. People argued that the money used to perform *kastom* for 'deceased' clan members was only available because of the hard work and the feasts previously performed by the 'deceased' themselves, which secured ownership of the very ground within the SML that now provides wealth for the current generation of landowners.

The Cultural Logics and Hazards of Exchange

These events remind us that mortuary feasts are never unproblematic sites of social reproduction, free from contest, critique or failure. *Kastom* is always threatened by backstage tensions within the core group of hosts, the potential for unforeseen problems (especially with last minute preparations), and the possibility that guests will not play their role properly, by not bringing promised pigs, refusing to eat the feasting food, or disturbing the proceedings through fighting or drunken behaviour — not to mention sorcery. In reality, *kastom* pivots on anything but virtuous sociality, which means that there is little unanimity or clarity over the meaning of *pasin bilong Lihir*. There is an obvious tension between dogma and praxis. Moreover, by putting their wealth into circulation and expending vast resources, hosts and allies risk the real possibility that people will fail to reciprocate in culturally (and economically) appropriate ways. Ultimately, *kastom* is 'dangerous circulation' (Foster 1993).

Ideally, exchange follows a typical cyclical pattern. As groups and individuals exchange items with one another, and hold feasts that nurture others, these processes are reciprocated, ensuring a sense of inclusion, mutual indebtedness and the 'repayment' of previous 'investments'. Traditionally, the objective was not to make a 'profit', or to return more than one receives. This is the basis for the current ideal of equality. In the past, when groups exchanged pigs for shell valuables and presented yams for consumption, these items were invested with the same 'cultural value'. The pigs were raised locally, yams were home-grown, and *mis* was not a mere currency, but represented the moral standing of individuals and groups. Older Lihirians often speak in terms of a closed cycle of reciprocal exchange of cultural equivalents. To an extent, isolation ensured that exchange was once very parochial, which may account for the development of strict notions of equivalence.

An equivalent gift should not only look similar to the original, but should ideally embody a similar level of physical effort, nurturance, and culturally significant skill, all of which are required for its creation, growth, exchange and eventual consumption. What is exchanged represents aspects of the donor; it is the moral autonomy invested in the gift that must be reciprocated. This is not to be confused with Maussian notions of *hau*, or an equivalent 'spirit of the gift' (Mauss 1925). As Sahlins (1974: 157) argues, we should avoid the temptation to try and understand the economic principle by concentrating on the 'religious' overtones of the concept. Godelier (1999: 106) presents a similar criticism, asserting that things do not move of their own accord, but only by the will of individuals. In debates surrounding the inalienability of gifts, it remains a moot point whether they are reciprocated because of claims over the produced object or simply the debt created through giving it to another person (Strathern 1982b: 549). Nevertheless, in Lihir the obligation is to return and perpetuate what is engendered through the act of giving. However, this becomes more problematic when we consider what is actually given in contemporary *kastom*.

As in other parts of New Ireland, Lihirian exchange and circulation are premised on a cultural logic of revelation and subsequent concealment. This parallel process permeates the wider social sphere, as individuals and groups elicit particular social values and responses from one another (Wagner 1986; Eves 1998). This enables people to regulate what others see and know about their identity, particularly their wealth (Foster 1993).[8] Richard Eves (2000) has extended this observation in order to explain the perceived rise in sorcery among the Lelet. He argues that the expansion in forms of wealth through their engagement with the market economy has brought a loss of control over the processes by which sociality was previously regulated. People are no longer able to effectively reveal and conceal wealth at will. Certain types of modern wealth — such as trucks or permanent housing materials — simply fail to fit existing

---

8   See Strathern (1985) for a more general discussion on the relationship between seeing and knowing throughout Melanesia.

cultural logics governing the categories of wealth and value. Such things cannot be hidden from view, nor is this is always desired. Individuals and groups are more vulnerable to sorcery that may arise through jealousy, envy or discontent over new forms of inequality. Thus the correlation between modernity and the discourse of sorcery is culturally constituted and not simply the generic by-product of global forces.

In Lihir, the sheer scale and variety of wealth, and the related inequality, surpass anything found in Lelet, and make Lihirian feasting and exchange a very risky business. The stakes have risen not only in terms of potential gains but also real and perceived losses. The rise in sorcery accusations that accompany times of upheaval or rapid social and economic change (or 'modernity') has been regularly noted in the developing countries of Africa (Geschiere 1997; Ferguson 1999, 2006; Meyer 1999). It therefore comes as no surprise to find that changes in Lihir have been followed by a definite increase in the discourse of sorcery (though not necessarily more incidents of the practice, which is rather more difficult to quantify). This is no doubt related to the double bind of exchange that has more general social implications: one must exchange wealth and hazard the dangers of circulation in order to produce social relations.

On Commodities, or What is a Lihirian Gift?

If exchange items signify personal investment, then how should we understand the transaction of 'gifts' and 'commodities' in these feasts? Melanesian exchange systems have mainly been understood through the intellectual legacies of Marcel Mauss. There has been an inordinate amount of anthropological attention paid to gifts at the expense of commodities, or simply an assumed agreement on the nature of commodities. While there has been some insightful work produced on the nature of commodities (Godelier 1977; Taussig 1980; Gregory 1982; Appadurai 1986; Miller 1987), we cannot assume that their character or constitution is axiomatic.

The terms of reference for anthropological understanding of the relationship between gifts and commodities were substantially influenced by the (1982) publication of Gregory's book, *Gifts and Commodities*. His work has been evaluated and challenged, but for the most part his distinction still provides the analytical template for this type of study. At the heart of his work lies a tidy formulation that is convincing and easy to remember: inalienable objects are exchanged as gifts between mutually dependent transactors, while alienable objects (commodities) move between mutually independent transactors, typically engaged in market relations. Gregory draws directly from the theoretical wellsprings of Mauss, arguing for an 'indissoluble bond' between the giver and the gift (inalienable possession), which is contrasted with (alienated) commodities to which people have no real connection. From this overdrawn opposition, it is usually inferred that the foundational principles of both forms of exchange are reflected in contrasting moral evaluations. As Gell (1992: 142) so neatly put it: Gift-Reciprocity-Good/Market-Exchange-Bad.

Although Gell pithily captures the essence of this overstatement, Sahlins and others before him first noted the intervals of sociality and morality in reciprocal forms of exchange that effectively place exchanges within a spectrum of reciprocities (Sahlins 1965).[9] Even if traditional exchange is generally characterised by morality and constraint, this does not imply everyone acquiesces in it, or that there are no contradictions, such as inclinations of self-interest in societies that customarily demand high levels of sociability (Sahlins 1974: 203). Variables such as kinship distance, wealth, rank, and the actual items of exchange inevitably determine the nature of reciprocity. Thus, what we find in the ideology of *kastom* and assertions of *pasin bilong Lihir* are fantasies of reciprocity that overlook or conceal these contradictions, variables and gradations, so that all economic interactions are 'judged' against an idealised 'pure gift'.

The actual performance of Lihirian *kastom* would appear to support Gell's unorthodox proposal that distinguishes between gifts and commodities, not on the basis of the character of the relationship between people and things (alienable/inalienable) or between people and people (independent/dependent), but rather the nature of the social context of a given transaction (Gell 1992: 146). This has particular significance for Lihir, where people exchange things that are hardly inalienable possessions, but nevertheless fulfil intended social purposes. Perhaps, then, it is useful to consider Apadurai's (1986) insights on the social life of things: at various stages or 'phases' in the life or 'career' of an object, it may be a commodity or it may not. In which case, gifts that are said to be 'inalienable' are not merely the inversion of commodities.

The production of commodities is a cultural and cognitive process. Commodities must be made as material items, and labour needs to have a price, but they also need to be culturally recognised as a certain type of thing. This echoes Simmel's (1978) observation that value is never an inherent property of objects, but is rather a judgment made about them by particular actors. Various things might be marked as commodities at one time and treated quite differently at another, and what one person recognises as a commodity may not be treated in the same way by someone else. Kopytoff (1986) refers to this as the 'cultural biography' of commodities. Things destined to be commodities can become something completely different, such as a gift. In order to comprehend Lihirian exchange,

---

9   See also: Malinowski 1961; Oliver 1955; Gouldner 1960.

we need to grasp the mutability of objects. If we view commoditisation as a process of 'becoming' rather than a 'terminal' category (an all-or-nothing state of being), then it is possible to emphasise the social context of exchange rather than the nature of the object itself.

So far, these Lihirian feasts are beginning to reveal more than they intended about the objects being exchanged and the people exchanging them. Aside from the interpenetration between *kastom* and the cash economy, we can see the disguised processes that re-value the goods in circulation. Concentrating on the context might allow for the possibility of substitution, but does this imply that the mere the act of exchange brings forth the desired social values? Put otherwise, could any token be used for the sake of eliciting an ideal sociality? Apparently not: it is not possible to use just anything. The ideology of *kastom* is premised upon the use of culturally recognised objects, and is rhetorically resistant to the use of certain goods. Despite the flexibility, there are still specific ideas about the constitution of an appropriate gift. So perhaps it is still necessary to focus on these things.

Gifts from the Ground and Gifts off the Shelf

Yams distributed during feasts are ultimately an objectification of the lineage and the clan's ability to enliven itself through productivity, their capacity to coordinate themselves underneath their big-man, and his own competence in making an event successful. In the social language of food, yams are an important idiom that expresses commonality between the donor and the receiver. In daily contexts, women cook yams for the household, and during feasts, men often cook yams in the men's house. The presentation of yams during feasts is part of the male performance of objectifying and distributing nurturance to clan members, allies and members of the opposite moiety. These actions extend the daily efforts of women in nurturing and sustaining individuals and lineages.

Presenting yams also represents the ability to produce healthy and abundant crops. In the past, garden magic was more central to agricultural production. Gardening involves a division of labour between males and females. Much of the strenuous work — such as clearing, fencing and tilling — is performed by males, while females are more engaged in planting, careful tending and harvesting. Males usually take the public credit for a successful feast (which implies the ability to produce plentiful crops), but they recognise their dependency upon female labour. Gifts of pork (*puatpes*) specifically acknowledge this contribution.

On mainland New Ireland, where taro is the staple feasting tuber, Clay recalls some instances in which Mandak people purchased taro, or hired other people to plant gardens for them after their own crops failed. However, purchasing taro for mortuary rituals 'was a mark against those sponsoring the feast ... [and]

buying taro with cash from copra and cacao sales added nothing to the buyer's prestige' (Clay 1986: 138). Eves (1998: 242) cites a similar reluctance among the Lelet to purchase taro for mortuary feasts. At the start of mine construction in 1995, Lihirians insisted that they would continue planting gardens for feasts and they would not include trade store goods in these events, indicating the cultural significance of food invested with the giver's time, substance, effort and knowledge. This is the way in which communal feasting enables groups to nurture and continue a gathered sociality. Before long, however, rice and other trade store goods came to be regarded as a substitute for garden produce. Lihirians in paid employment, supposedly without the time to maintain large gardens, were expected to contribute to feasts by purchasing and presenting trade store items. This also expressed the prestige associated with having access to money within the clan. The use of store food was initially justified as an appropriate way for workers to 'repay' the efforts of parents and relatives who previously paid for school fees, provided food and nurturance, and made other beneficial purchases throughout their earlier lives. Over time, this has evolved into a general expectation for their inclusion in *kastom*. Nevertheless, despite these changes, disdain for the use of purchased produce for ritual purposes remains central to the ideology of *kastom*.

Porcine Presents

We find similar contradictions in the transaction of pigs. Nostalgia and desire for the exclusive exchange of locally domiciled pigs is set against the common mainland New Ireland practice of purchasing pigs from other areas. The use of pigs from elsewhere is a concealed statement about the contradictory nature of status. Big-men should be free of debts, yet the display of pig debts in mortuary feasting reflects the prestige which they have gained through links outside the clan and beyond Lihirian shores — not to mention the ability to purchase pigs with cash.

As in other parts of Melanesia, pigs are significant for their pre-eminent value as food for human consumption, as the dominant object of exchange, and as repositories of tremendous symbolic potency (see Jolly 1994: 173). Lihirians take great pride in raising their pigs, and recognise the close relationship identity between pigs and humans.[10] Lihirians describe feeding a pig as *tananie a bual*, which is another form of the verb *tinen*, which carries the sense of caring, nurturing or feeding people. Pigs are not simply the embodiment of labour and

---

10   The close bond between humans and pigs, and the significance of pigs in various aspects of social life, is a general Melanesian phenomenon (see Bulmer 1967: 20). They are central items in exchange cycles (Strathern 1971b; Megitt 1973), victims of ritual slaughter (Rappaport 1968; Keesing 1976), and a common feature in bridewealth payments and war compensation, where they can be regarded as 'substitutes' for human life (Glasse and Meggit 1969; Modjeska 1982; Macintyre 1984).

productivity. Older Lihirians claim specialist knowledge or 'talk' that makes their pigs grow faster and bigger. This knowledge is in decline among younger generations, but the capacity to raise large healthy pigs is still considered to be the result of hard work and the ability to deploy specialist knowledge in productive ways. Pigs are also a primary means for reconstituting damaged social relationships. Pigs are identified with the donor, and consumption within the men's house rectifies an impaired moral relationship. The presentation of live pigs for exchange and cooked pork for consumption is an objectification of the moral attributes of the donor. They are not singular isolated gestures; they predicate a similar response from the receiver at a later date. These presentations convey a complex message of moral autonomy which represents the donor's capacity for autonomous action in relationship to someone who is 'deceased'. Here morality involves individual self-determined action directed towards the nurturance of relationships.

In reality, since mining operations began, fewer Lihirians are rearing pigs, despite the fact that some people have started small piggery projects (*banis pik*). The stated preference for locally domesticated pigs is a way of presenting Lihirian pride that has little relation to what is actually happening, highlighting the ambivalent character of pigs as practical necessities and symbolic repositories. The purchase of foreign pigs is a way for men to display their affluence and put others in debt, for reciprocation requires a pig of equivalent size and a similar level of expenditure. On any given day, people from the neighbouring islands of Tanga and Tabar arrive in dinghies in Lihir to sell their pigs. These are not always prearranged customary transactions: other New Irelanders have seized upon the opportunity to exploit a flourishing 'exchange market'. Pigs are valuable for their association with *kastom* and social reproduction, but also because one pig can fetch over K4000. People from other areas have complained of the internal disruption this has brought to their own exchange relations. Some people are disinclined to trade pigs within their local networks unless people can match the price paid by Lihirians. In the absence of other avenues for making money, pigs are an important source of cash which is then redirected back into other local exchange systems. Effectively, the commoditisation of pigs for Lihirian *kastom* has become the means for other communities to engage with the cash economy and modern forms of consumption.

Social Currency

The contemporary ceremonial economy does not revolve around the exchange of inalienable possessions. This is not to say that there are no highly ranked shell valuables that must remain within the clan, where ownership reflects political standing. However, by 2000 or even earlier, certain types of highly prized shell money, such as the rare *pangpang* or the *ndolar* arm shells, were rarely active in the local economy.[11] Gifts are supposed to embody the 'spirit' and substance of the giver (or the producer), but there is no 'insoluble bond' that compels reciprocity. The absence of substance does not necessarily detract from the outward significance of the gift or the sociality created through exchange, yet this fact is complicated by the ambiguous nature of *mis* and other shell valuables. Like pigs and yams, they fulfil multiple purposes and possess a range of values. *Mis* has proven equally fluid: it is simultaneously a prized valuable, a daily currency, and a commodity.

According to Wagner, the Barok counterpart (*mangin*) approximates the functions of three kinds of currency: it is 'vital wealth', as part of the economy of human attachments; it is 'money', as an exchange medium in the local economy and as a good that can be redeemed by state money; and it is 'money' in another sense, as a medium of moral merit (Wagner 1986: 83). *Mis* can be understood within this framework, for like *mangin*, 'it approximates a kind of triple metaphor: each standard of exchange draws away from the definitional certitude of the other two' (ibid.). As in Godelier's (1977) description of Baruya salt currency, the character of *mis* shifts on the basis of the social context of its use.

Lihirians proclaim the value of *mis* in highly moral terms. It is reified as the pecuniary analogue of the 'stable' and 'social' nature of Lihirian society juxtaposed against an essentialised, anti-social, cash-based society (see Bloch and Parry 1989: 6). However, notional opposition stems also from an ongoing struggle for representation of self and other which has only intensified in the context of mining, as Lihirians have willingly adopted contrasting orientalisms and occidentalisms about cash and shell money. In reality, antagonism is not directed at cash *per se*, but at the manner in which it so unequally distributed in the context of mining. People are thus compelled to the performance of *kastom* as a means for redistributing wealth and achieving unity. Many people recognise a lack of 'control' over *kastom*. This seems to prompt them to imagine that, because *kastom* is Lihirian, they should be able to 'control' it and keep it separate from the cash economy. This is especially true of the Society Reform leaders, the LJNC, and the authors of the Destiny Plan. The Destiny Plan proposes strategies to subject the cash economy to Lihirian interests, ideally to regulate *kastom*, *bisnis*, and the corporate mining economy. So *mis*, as the money of *kastom*, partly retains value or significance through its *ideological* insulation from the cash economy. It is valuable because it is *described* as incommensurate

---

11   These precious items are considered priceless and rest in the hands of a few elderly men. As elderly men feel less tolerant of an increasingly 'disrespectful' younger generation of men, they are disposing of this wealth, sometimes by burying it or throwing it out to sea, to ensure that these younger men are not the beneficiaries of hard work and high status that are no longer appreciated. At the same time, some men are hoarding their shell valuables and thus reducing their use value.

and should be used for certain transactions with specific social purposes. But in such radically different circumstances, the enduring value is also contingent upon the wide variety of uses to which *mis* can be put.

Since mining began, the production of *mis* has greatly increased. When it is manufactured for sale, it does not acquire cultural significance until it is used in *kastom*. Otherwise it is obtained through social relationships. The majority of Lihirian *mis* are not inalienable possessions. While some shell valuables which are 'loaned' or transacted must eventually return, in most cases there is no intention to recover the same strand. More often than not, *mis* functions as general purpose currency that is commensurate with various objects and services.

Mis is nearly always used with the intention to establish or maintain relationships, to correct a social imbalance, or provide someone with the means for entering into new relationships: it is a key item in the process of social reproduction. It derives part of its value from being an exchange item that binds people together in different ways, which gives each strand an important social history. But Lihirians happily use *mis* for a variety of needs outside of ceremonial exchange, such as for decoration in church celebrations or school graduations, or as gifts to visitors — which may indicate the variety of activities now encompassed under the ideology of *kastom*. Alternatively, when it is used for the opening of new office buildings, businesses or trade stores, it plays a valuable role in the performance of the ideology of *bisnis*. Not surprisingly, many people complain that *mis* is now 'devalued' and has 'no meaning'. This reflects the paradox of meaning or the fantasy of value: people want to insulate or sequester *mis* and other items to increase their moral value (as the money of *kastom*), but at the same time, they want to use *mis* as a currency or for other uses, to reinforce its 'token' value and its Lihirian identity. Thus, both *kastom* and *mis* retain their significance through *practice* and *use*, even if those practices or uses are now rather different from before or contradict dominant ideologies.

This process is certainly attributable to the florescence of shell valuables that corresponds to the loosening of restraints on ownership and technical knowledge. *Mis* is no longer held solely in the hands of leading big-men; it is easily attained by men and women with money, regardless of their customary influence. If men are short of *mis* to purchase pigs for an upcoming feast, or caught out by an unexpected death, they can make arrangements to purchase more. When value is contingent upon scarcity, increased production ensures greater use, but it also facilitates the process of 'devaluation'. This is analogous to the devaluation of the pearlshells once used in Hagen *moka* transactions that were eventually replaced by money because of the massive influx of shells during the colonial period (Strathern 1971a, 1979; Hughes 1978; Healy 1985). But in this case, *mis* has not been abandoned for another exchange medium, and it continues to coexist with cash in nearly all *kastom* transactions. However, at a deeper level, the modern production of *mis* reveals the dependency of *kastom* on *bisnis*. For those people involved in the production of *mis*, it is explicitly the *bisnis* of keeping *kastom* alive, generating sociality through the medium of mortuary ritual.

Splicing the System

The persistence of mortuary rituals is less a consequence of any attempts by different groups to isolate these practices from outside influence, and more a result of the inherent flexibility and general capacity of Melanesian exchange systems to appropriate introduced goods. This point serves to highlight the contradiction within the ideology of *kastom*, which purports that insulation ensures continuity. This contradiction is thrown into further relief by the appeal to *kastom* as a means to achieve social stability, which demands the incorporation of new items and forms of wealth for wider distribution. Thus, the endurance of *kastom* also results from a historical preoccupation with achieving unity in the face of modern social divisions. The significant amount of time that Lihirians now spend memorialising their dead and enacting their ideology of *kastom* must be attributed to the more unpleasant aspects of modernity associated with mining.

The Lihir exchange system is regularly spliced with the capitalist economy. The purchase of goods for *kastom* represents a more general Melanesian phenomenon: modern Melanesians taming or domesticating commodities as they convert them from market goods into cultural gifts (see Weiner 1976: 78–9; Heaney 1982: 229; Carrier and Carrier 1989; Akin 1999). The ideal cyclical image, in which sociality is made to 'come up' through the reciprocation of 'equivalents', is maintained through customary practices that present certain goods as if they embody the culturally valued attributes that were traditionally invested in gifts. When Zipzip and the others purchased pigs from Tabar and other places, they usurped female contributions to *kastom* through their access to cash, removing female productive capacities from the equation, which effectively reduced the distinction between producers and transactors. Every time that Lihirians use commodities in *kastom* exchanges — be they pigs, shell money, garden produce from the market, or trade store food — their action sustains the cyclical image of continuity, but in reality, this 'cycle' is cut and spliced with the cash economy. The processes of develop*man* reinvigorate, vitalise and authenticate *kastom* by maintaining the illusion of continuity.

Lihirians are unlikely to exclude new goods from *kastom,* because there is a growing expectation for them to perform modernity within a locally defined cultural framework. Each *kastom* event increases the inertia of this consumptive pattern. The net result is the transformation of a system of delayed reciprocal exchange to the point where transactions that would have previously been delayed are now 'sped up'. While exchange functioned as a levelling device in the past, providing an important avenue for dispersing wealth and resources and consolidating group membership, the scale has now increased and form of wealth has changed. The exchange system previously entailed an ebb and flow in which wealth and prestige were somewhat 'regulated' by human productivity. The huge influx of money has destabilised the cyclical nature of exchange as exchange wealth grows within the hands of a minority. Mortuary rituals are now literally performed on unequal ground: the inequitable distribution of wealth means that there is no longer any equilibrium underpinning the pattern of exchange, and some people will stay in the game with far greater ease than others.

We might conclude that Lihirian society has been 'reformed' by the global forces of capitalism exerted through the mining operation, and that *kastom* — or the local moral economy — is being hollowed out in theory and in practice. One inference that we can make from the processes embedded within these feasts is that cultural continuity is an illusion. Some Lihirians would certainly agree, and this strengthens their resolve for a return to true Lihirian ways and their commitment to the ideology of *kastom*. But we also have to recognise that many of these changes have arisen internally and intentionally to ensure the continued practice and relevance of *kastom* in entirely altered circumstances. While this might not happen in the ways proposed by the authors of the Society Reform Program or the Destiny Plan, Lihirian mortuary rites are self-consciously performed as *kastom* — as activities that are both indigenous and ancestral, which is the source of their cultural authenticity and validity.

8. Conclusion: Society Reformed

Having traced the transformations that have taken place over the past century, it is obvious that Lihirians are not simply advancing towards a final destination in an inevitable world-historical teleology. They are actively shaping their lives and the forces that impinge on their existence. They are using new things and opportunities for their own purposes, although often with unforeseen outcomes. Lihirian society has been irreversibly reformed, but Lihirians have never passively capitulated before the global capitalist system, nor does everyone regard all of the changes as entirely unwanted. While there is a definite nostalgia for an idealised past, exacerbated by a strong sense of cultural rupture coupled with a hyper-traditionalism, nobody imagines a utopian return to primordial life and ancestral ways. Traditional Lihirian culture might have had superior values, but money, trade store food, televisions, beer and cars were not part of that era. Modern Lihirian culturalism is premised on the demand for new things, or more precisely the requirement to indigenise them. In a veritable hall of mirrors, Lihirian desires reflect and refract Western dreams of an advanced urban egalitarian society where everybody has the capacity for endless consumption in order to advance their ideas about what life is all about.

At the same time, Lihirians remain divided over their hopes for modern life and how they are best achieved. Lihirians are highly aware that the new influences, challenges and agendas brought by large-scale resource development have created deep social divisions. The tremendous form of change created through mining means that there is little unison over many aspects of their lives, such as leadership, the use and inheritance of resources and wealth, social values, gender roles, governance, which road will lead to the imagined future, and even what this new life will look like. However, in response to the extreme experience of mining and the lack of consensus over so much of their life, it seems that many Lihirians look to *kastom* as the one thing that they should all be able to agree upon. As a result, their efforts are doubled in this direction in order to create and maintain a semblance of social and cultural unity and continuity.

Lihirians might argue that things were better in the good old days, but with the arrival of the mine, their ancestors have never been venerated with such style, colour and magnitude. Lihirians have proven themselves to be quick students of commercial cunning — especially as they craftily extract concessions from the mining company — and they use this to stage the most elaborate traditional feasts they have ever known. Lihirian economic practices amply demonstrate that, although Western capitalism is planetary in scope, it is not a universal logic of cultural change (Sahlins 2005c: 495). Lihirian inventions and inversions of tradition are their attempt to create a differentiated cultural space within the world system — effectively the Lihirianisation of modernity. As Joel Robbins (2005: 10) points out, contemporary customary activities should be seen as a new form of cultural process, not symptoms of the 'death of culture' or its incoherence or irrelevance.

Contrary to popular Western thought, the introduction and use of new goods has not meant that Lihirians had to adopt the whole cultural package that accompanies money and other useful items, nor has it necessarily set them on an inevitable course of cultural corrosion that will eventually rust out any trace of their traditional existence. As Thomas (1991: 186) notes, this line of thought reiterates the spurious assumption that 'material culture is an index of acculturation'. However, if Lihirian strategies have allowed *kastom* to be carried forward into a new world, this has not been without cost. As we have seen, the practices and values associated with mortuary rituals have been transformed in ways that people find profoundly disturbing. Moreover, it would appear that the particular version of *kastom* which Lihirians practice seems to divide them even further — and this is the paradox behind the ideology of *kastom*. Even so, many Lihirians remain convinced that *kastom* is the true road that will lead them (back) to a state of equality and social balance — but in a world where everyone is also equally rich. Many Lihirians have been left baffled as to why this conviction has proven tragically false. This is a familiar story throughout all of Melanesia's resource development projects, where unprecedented amounts of cash and resources injected into communities routinely undermine even the most flexible social systems of leadership and distribution. This may well confirm that *kastom* was simply never equipped to deal with mineral wealth.

In response to these dilemmas, the authors of the Destiny Plan have proposed a way out of the quagmire which they believe will allow Lihirians to achieve the dream of a reformed, egalitarian and well-off society — to live fuller *Lihirian* lives. However, the Lihir Destiny, as a concept and a destination, remains ambiguous, if not downright contentious. This is partly because people cannot agree upon which road will lead them there, but also because these roads seem to point to different visions of how this life will be lived.

At one level, Lihirians are caught between competing cultural values seemingly ascribed to different activities and ways of being. But even though the Destiny Plan aims to modernise Lihir according to perceived Western values — to hasten the transition from develop*man* to development — it is not premised upon, nor has it generated, the kind of cultural humiliation which Sahlins suggests is necessary for the total abandonment of Lihirian culture. Instead, we can see the double-edged side of humiliation, which has created a heightened self-consciousness and spurred on a greater commitment to retain a distinctive Lihirian identity. The Destiny Plan might be structured around the philosophies of Personal Viability that demand a refashioning of the self, but in the imagined future, *kastom*, kinship and traditional values and epistemologies remain central, albeit in a highly regulated form. The question is not whether these aspects of the Lihirian lifeworld should be discarded, but rather how they can be managed so that people can simultaneously recognise themselves as Lihirian and live modern lives in a developed society. In the same way that we might consider more classical cargo cults, or perhaps even the Nimamar movement, as an attempt to transcend the binary opposition of develop*man* and development in order to achieve something entirely new, a creative synthesis, there are some ways in which the Destiny Plan pursues a similar ambition. This analogy should also alert local leaders, mining company management and the government to the limitations or some of the internal contradictions in this vision.

The Destiny Plan is not a simple road to modernity, partly because it encompasses the hybrid cultural, economic and ideological space in which many Lihirians find themselves. While the Destiny Plan indicates that the key to long-term economic independence — indeed the future viability of Lihirian society — can be found in entrepreneurialism, it emphasises *bisnis* in the narrowest of terms, overlooking or concealing the fact that many Lihirians also use the term *bisnis* to refer to relationships that are fostered through *kastom*. This strategy ultimately fails to recognise the absolute entanglement between these spheres, or the level to which both of these activities currently rely upon the corporate mining economy. Needless to say, this ideological distinction is routinely undermined by the ways in which Lihirians engage with capitalism as they pursue the develop*man* project. Moreover, even though the authors of the Destiny Plan reject the so-called cargo mentality and unrealistic landowner expectations, as expressed through local manifestations of the dependency syndrome, they still expect an 'uncalculating gift' from the mining company (Godelier 1999: 208). Despite their criticisms of landowner proclivities and their protestations about the need for self-reliance, the ideology of landownership has a complete stranglehold on them. Thus we might consider the Destiny Plan as a simulacrum that echoes the outward manifestation of the modern world, with its incessant categorisation, hierarchies, distinctions and control. From this perspective, we begin to see the centrality of mimesis as practice, and the enduring tension between knowledge and implementation which ultimately frustrates the immediate realisation of the Lihir Destiny.

The widespread expectation that the company will deliver all forms of economic development means that many Lihirians, and especially landowners, fail to see the need for PV. Ultimately, mining benefits enable landowners to live Lihirian lives on a bigger and better scale; they are not totally reliant on farming, nor do they have to front up to the market place for the expected hiding. Their subsistence existence is augmented by their freedom to consume. As mining benefits subsidise the ceremonial economy, people are able to pursue develop*man* with more splendour and pageantry and boost their own political standing at the same time. The majority of non-landowners might resent the landowners for appearing selfish, for failing to redistribute wealth in expected ways, or for polarising their political status, but this barely detracts from their desire to enjoy and partake in such an existence.

Nevertheless, emergent class relations linked to the arbitrary allocation of landowner wealth and status, coupled with the daily reminder of global inequalities reflected in the wealth of the mining company and its expatriate managers, has left some people susceptible to PV promises. The same capitalist system that has consistently denied Lihirians equal footing with their colonial *masta*, their proselytising missionaries, their expatriate bosses, and their adopted anthropologists, has been repackaged and sold back to them as something new that will enhance rather than detract from their lives. Through a conceptual sleight of hand, capitalism is presented as a fair system that is able to increase, rather than decrease, social equality, and as one that simultaneously requires self-regimentation and offers hitherto unimagined possibilities for personal accumulation. The morality of the capitalist system depicts differences between the rich and the poor, or elites and grassroots, as differences in degree rather than differences in kind. In the words of Gewertz and Errington (1999: 42), such differences come to reflect 'a relatively fluid continuum of personal attributes rather than a relatively closed set of categorical differences'. Economic inequalities are thus not only fair but necessary, because they represent people's contribution to society and the efforts they invest in the processes of production. In this new world, the individual is responsible for community well-being. As Polanyi (2001: 114) would have it, the assumption is similar to Mandeville's 'famous doggerel about the sophisticated bees' who demonstrated how private vice can yield public benefit.

The strategies outlined in the Destiny Plan require people to be *interacting* individuals — or 'dividuals' as Marilyn Strathern might see it. Depending on the context or the task at hand, they might act (or imagine themselves) as sole proprietors of the self in a *bisnis* transaction, and then acknowledge their relationally embedded position in society when performing *kastom*. But in reality, it is not always so easy to make an effortless transition or to maintain the boundaries between supposedly different economic spheres, nor does everyone necessarily want to engage in such cultural acrobatics. For example, when I once asked Francis Bek why his small entrepreneurial endeavours had not succeeded (why his PV 'money garden' failed to take root), he replied that it was simply because 'the ways of Lihir' were too fixed in him (*pasin bilong Lihir em i pas pinis long mi*). Recognising this 'embodiment of history' (Bourdieu 1977) helps us to understand the field of expectations which many Lihirians are trying to negotiate their way through.

But there are some people, such as members of the emerging elite, who successfully manage their demanding relationships, as well as a growing number of people who wholeheartedly embrace PV as a strategy to achieve such outcomes. Samuel Tam and the authors of the Destiny Plan have been seeking to create an environment in which people can constitute themselves as possessive individuals. Indeed, as Hobbes would have it, in the world imagined by PV, it is possessive individuals pitted against one another all the way down. There can be no other way. If PV is appealing, it is perhaps because it articulates what people already suspected: that private ownership, management and consumption of wealth underpins the sort of lifestyles being presented as genuinely modern. The strong identification between modernity and certain forms of consumption and ownership means that possessive individualism becomes something worth striving for. For some, PV taps into an incipient desire to break away and distinguish one's self from people embedded in tradition, collective obligation, and consequent relative economic poverty — traits of 'backwardness' in the over-extended distinction between tradition and modernity.

When elite leaders like Mark Soipang leave their air conditioned office in Londolovit and drive to Putput village to exchange their suit for traditional attire and self-consciously mount the stage to perform the *rohriahat* rituals in the final *karat* feast, as Soipang did in 2008, they demonstrate that they are neither fully beholden to tradition or to modernity. These 'masters of development', who are still deeply involved in develop*man* projects, reveal the room for creative action that has been opened up in these new circumstances. These leaders are not still wandering through the desert of cultural humiliation in search of the other side. Instead, their activities tell us that cultural transformation does not occur on a linear scale, but appears as people negotiate their way between idealised and supposedly opposed states. What I have presented in this book are the tensions that exist as people move between develop*man* and development and negotiate a new hybrid space: the cancellation of essentialised difference through imaginative synthesis.

However, it is worth remembering that the develop*man* process contains a historical and structural paradox which Lihirians cannot avoid. As long as Lihirians continue to equip themselves with fancy and useful things from the market for the vitalisation, reproduction and progression of their own cultural order, their culture will become increasingly dependent upon the relations of the world system as it is manifest locally through the business of resource extraction. This is surely a point which the authors of the Destiny Plan have recognised. Unfortunately, in such a hostile context of global capital, and with the inevitable prospect of mine closure, there are particular types of develop*man* that may prove devastatingly self-destructive. Perhaps only then will Lihirians experience total humiliation. But it remains to be seen whether this will make them truly modern subjects of the global capitalist order, or whether they will lose their ideologies and fantasies but keep their customary practices.

References

Akin, D., 1999. 'Cash and Shell Money in Kwaio, Solomon Islands.' In D. Akin and J. Robbins (eds), *Money and Modernity: State and Local Currencies in Melanesia*. Pittsburgh: University of Pittsburgh Press.

——, 2005. 'Ancestral Vigilance and the Corrective Conscience in Kwaio: Kastom as Culture in a Melanesian Society.' In J. Robbins and H. Wardlow (eds), *The Making of Global and Local Modernities in Melanesia: Humiliation and the Nature of Cultural Change*. Aldershot: Ashgate Publishing.

Allen, M., 1984, 'Elders, Chiefs, and Big Men: Authority, Legitimation, and Political Evolution in Melanesia.' *American Ethnologist* 11: 20–41.

Amarshi, A., K. Good and R. Mortimer, 1979. *Development and Dependency: The Political Economy of Papua New Guinea*. Melbourne, Oxford University Press.

Appadurai, A., 1986. 'Introduction: Commodities and the Politics of Value.' In A. Appadurai (ed.), *The Social Life of Things: Commodities in Cultural Perspective*. Cambridge: Cambridge University Press.

Awart, S., 1996. 'Women of Lihir: Coping with Culture Change.' *Research in Melanesia* 20: 1–36.

Bainton, N.A., 2006. Virtuous Sociality and Other Fantasies: Pursuing Mining, Capital and Cultural Continuity in Lihir, Papua New Guinea. Melbourne: University of Melbourne (Ph.D. thesis).

——, 2008. 'Men of Kastom and the Customs of Men: Status, Legitimacy and Persistent Values in Lihir, Papua New Guinea.' *Australian Journal of Anthropology* 19: 195–213.

——, 2009. 'Keeping the Network Out of View: Mining, Distinctions and Exclusion in Melanesia.' *Oceania* 79: 18–33.

Bainton, N.A., C. Ballard and K. Gillespie (forthcoming a). 'The Beginning of the End: Sacred Geographies, Memory and Performance in Lihir.' *The Australian Journal of Anthropology*.

Bainton, N.A., C. Ballard, K. Gillespie and N. Hall (forthcoming b). 'Stepping Stones Across the Lihir Islands: Developing Cultural Heritage Management Strategies in the Context of a Gold Mine.' *International Journal of Cultural Property*.

Ballard, C. and G. Banks, 2003. 'Resource Wars: The Anthropology of Mining.' *Annual Review of Anthropology* 32: 297–313.

Banks, G., 1996. 'Compensation for Mining: Benefit or Time-bomb? The Porgera Gold Mine.' In R. Howitt, J. Connell and P. Hirsch (eds), op. cit.

——, 1998. 'Compensation for Communities Affected by Mining and Oil Developments in Melanesia.' *Malaysian Journal of Tropical Geography* 29: 53–67.

——, 2002. 'Mining and the Environment in Melanesia: Contemporary Debates Reviewed.' *Contemporary Pacific* 14: 39–67.

——, 2006. 'Mining, Social Change and Corporate Social Responsibility: Drawing lines in the Papua New Guinea Mud.' In S. Firth (ed.), *Globalisation and Governance in the Pacific Islands*. Canberra: ANU E Press.

Banks, G. and C. Ballard (eds), 1997. *The Ok Tedi Settlement: Issues, Outcomes and Implications*. Canberra: National Centre for Development Studies (Pacific Policy Paper 27).

Bashkow, I., 2006. *The Meaning of Whitemen: Race and Modernity in the Orokaiva Cultural World*. Chicago: University of Chicago Press.

Bell, F.L.S., 1934. 'Report on Field Work in Tanga.' *Oceania* 4: 290–310.

——, 1935a. 'The Avoidance Situation in Tanga.' *Oceania* 6: 175–98.

——, 1935b. 'Warfare among the Tanga.' *Oceania* 5: 253–79.

——, 1937. 'Death in Tanga.' *Oceania* 7: 316–39.

——, 1938. 'Courtship and Marriage among the Tanga.' *Oceania* 8: 403–18.

——, 1947. 'The Place of Food in the Social Life of the Tanga.' *Oceania* 17: 10–26.

Benyon, R., 1996. Socio-economic Impact of a Gold Mine on the Island of Lihir, Papua New Guinea: Land at the Heart of the Matter. Noumea: French University of the Pacific, Noumea Campus (Pre-Doctoral Diploma thesis).

Biersack, A., 1999. 'Porgera ⊠ Whence and Whither?' In C. Filer (ed.), op. cit.

Billings, D., 1969. 'The Johnson Cult of New Hanover.' *Oceania* 40: 13–9.

——, 2002. *Cargo Cult as Theatre: Political Performance in the Pacific*. Lanham (MD): Lexington Books.

——, 2007. 'New Ireland Malanggan Art: A Quest for Meaning.' *Oceania* 77: 257–85.

Bloch, M. and J. Parry, 1989. 'Introduction: Money and the Morality of Exchange.' In J. Parry (ed.), *Money and the Morality of Exchange*. Cambridge: Cambridge University Press.

Bourdieu, P., 1977. *Outline of a Theory of Practice*. Cambridge: Cambridge University Press.

——, 1984. *Distinction: A Social Critique of the Judgement of Taste*. Cambridge: Harvard University Press.

Bulmer, R., 1967. 'Why is the Cassowary Not a Bird: A Problem of Zoological Taxonomy among the Karam of the New Guinea Highlands.' *Man* 2: 5–25.

Burley, A., 2010. Ecological Impacts of Socio-Economic Change on a Tropical Forest: The Demography of Traditional Forest Resource Tree Species on Lihir Island, Papua New Guinea. Melbourne: University of Melbourne (Ph.D. thesis).

Burridge, K., 1971. *New Heaven, New Earth: A Study of Millenarian Activities*. Oxford: Basil Blackwell.

——, 1995 [1960]. *Mambu: a Melanesian Millennium*. Princeton: Princeton University Press.

Burton, J., 1997. 'The Principles of Compensation in the Mining Industry.' In S. Toft (ed.), op. cit.

——, 1999. 'Evidence of the "New Competencies"?' In C. Filer (ed.), op. cit.

Callister, G., 2008. 'Illegal Miner Study: Report on Findings.' Unpublished report to the Porgera Environmental Advisory Komiti.

Callister, S., 2000. A Cord of Three Strands Is Not Easily Broken: Birth, Death and Marriage in a Massim Society. Sydney: Macquarie University (Masters thesis).

Carrier, J. and A.H. Carrier, 1989. *Wage, Trade, and Exchange in Melanesia: A Manus Society in the Modern State*. Berkeley: University of California Press.

Casey, E., 1996. 'How to Get from Space to Place in a Fairly Short Stretch of Time: Phenomenological Prolegomena.' In S. Feld and K. Basso (eds), *Senses of Place*. Santa Fe (NM): School of American Research Press.

Chowning, A., 1974. 'Disputing in Two West New Britain Societies.' In A.L. Epstein (ed.), *Contention and Dispute: Aspects of Law and Social Control in Melanesia*. Canberra: Australian National University Press.

Clark, J., 2000. *Steel to Stone: A Chronicle of Colonialism in the Southern Highlands of Papua New Guinea*. Oxford: Oxford University Press.

Clay, B.J., 1977. *Pinikindu*. Chicago: University of Chicago Press.

——, 1986. *Mandak Realities: Person and Power in Central New Ireland*. New Brunswick (NJ): Rutgers University Press.

Comaroff, J., 1985. *Body of Power, Spirit of Resistance: The Culture and History of a South African People*. Chicago: University of Chicago Press.

Connell, J. and R. Howitt (eds), 1991. *Mining and Indigenous Peoples in Australasia*. Sydney: Sydney University Press.

Counts, D., 1971. 'Cargo or Council: Two Approaches to Development in Northwest New Britain.' *Oceania* 41: 228–97.

Crocombe, R., 2007. *Asia in the Pacific: Replacing the West*. Suva: IPS Publications.

Dalton, D., 2000. 'Introduction.' *Oceania* 70: 285–93.

——, 2004. 'Cargo and Cult: The Mimetic Critique of Capitalist Culture.' In H. Jebens (ed.), *Cargo, Cult, and Culture Critique*. Honolulu: University of Hawai'i Press.

Damon, F., 1990. *From Muyuw to the Trobriands: Transformations Along the Northern Side of the Kula Ring*. Tucson: University of Arizona Press.

Demaitre, E., 1936. *New Guinea Gold: Cannibals and Gold-Seekers in New Guinea*. Boston: Hougton Mifflin Company.

Denoon, D., 2000. *Getting Under the Skin: The Bougainville Copper Agreement and the Creation of the Panguna Mine*. Melbourne: Melbourne University Press.

Dobb, M., 1972 [1937]. *Political Economy and Capitalism: Some Essays in Economic Tradition*. London: Routledge & Kegan Paul.

Dwyer, P.D. and M. Minnegal, 1998. 'Waiting for Company: Ethos and Environment among Kubo of Papua New Guinea.' *Journal of the Royal Anthropological Institute* (N.S.) 4: 23–42.

Epstein, A.L., 1963a. 'The Economy of Modern Matupit: Continuity and Change on the Gazelle Peninsula.' *Oceania* 33: 182–214.

——, 1963b. 'Tambu, a Primitive Shell Money.' *Discovery* 24: 28–32.

——, 1969. *Matupit*. Berkeley: University of California Press.

Epstein, T.S., 1968. *Capitalism, Primitive and Modern: Some Aspects of Tolai Economic Growth*. Canberra: Australian National University Press.

Ernst, T., 1999. 'Land, Stories, and Resources: Discourse and Entification in Onabasulu Modernity.' *American Anthropologist* 101: 88–98.

Errington, F., 1974. 'Indigenous Ideas of Order, Time, and Transition in a New Guinea Cargo Movement.' *American Ethnologist* 1: 255–67.

Errington, F. and D. Gewertz, 1995. *Articulating Change in the "Last Unknown"*. Boulder (CO): Westview Press.

——, 2004. *Yali's Question: Sugar, Culture, and History*. Chicago: University of Chicago Press.

Eves, R., 1998. *The Magical Body: Power, Fame and Meaning in a Melanesian Society*. Amsterdam: Harwood Academic Press.

——, 2000. 'Sorcery's the Curse: Modernity, Envy and the Flow of Sociality in a Melanesian Society.' *Journal of the Royal Anthropological Institute* (N.S.) 6: 453–68.

Ferguson, J., 1999. *Expectations of Modernity: Myth and Meanings of Urban Life on the Zambian Copperbelt*. Berkeley: University of California Press.

——, 2006. *Global Shadows: Africa in the Neoliberal World Order*. Durham (NC): Duke University Press.

Filer, C., 1988. 'Report of Visit to Panguna with Lihir Landowners, 2/11/88.' Unpublished report to Kennecott Explorations Australia.

——, 1990. 'The Bougainville Rebellion, the Mining Industry and the Process of Social Disintegration in Papua New Guinea.' *Canberra Anthropology* 13: 1–39.

——, 1992a. 'The Lihir Hamlet Hausboi Survey: Interim Report.' Waigani: Unisearch PNG Pty Ltd (for Kennecott Explorations Australia and PNG Department of Mining and Petroleum).

——, 1992b. 'Lihir Project Social Impact Mitigation: Issues and Approaches.' Unpublished report to PNG Department of Environment and Conservation.

——, 1994. 'Sosel Impakt Bilong Lihir Gol Main: Ripot long Pipal Bilong Lihir.' Unpublished report to Lihir Mining Area Landowners Association.

——, 1995. 'Participation, Governance and Social Impact: The Planning of the Lihir Gold Mine.' In D. Denoon (ed.), *Mining and Mineral Resource Policy Issues in Asia-Pacific: Prospects for the 21st Century*. Canberra: Australian National University, Research School of Pacific and Asian Studies, Division of Pacific and Asian History.

——, 1997a. 'Compensation, Rent and Power in Papua New Guinea.' In S. Toft (ed.), op. cit.

——, 1997b. 'Resource Rents: Distribution and Sustainability.' In I. Temu (ed.), *Papua New Guinea: A 20/20 Vision*. Canberra: Australian National University, National Centre for Development Studies (Pacific Policy Paper 20).

——, 1998. 'The Melanesian Way of Menacing the Mining Industry.' In L. Zimmer-Tamakoshi (ed.), op. cit.

——, (ed.), 1999. *Dilemmas of Development: The Social and Economic Impact of the Porgera Gold Mine 1989–1994*. Canberra: Asia Pacific Press (Pacific Policy Paper 34). Boroko: National Research Institute (Special Publication 24).

——, 2004. 'Horses for Courses: Special Purpose Authorities and Local-Level Governance in Papua New Guinea.' Canberra: Australian National University, Research School of Pacific and Asian Studies, State, Society and Governance in Melanesia Project (Discussion Paper 2004/6).

——, 2006. 'Custom, Law and Ideology in Papua New Guinea.' *Asia Pacific Journal of Anthropology* 7: 65–84.

——, 2008. 'Development Forum in Papua New Guinea: Upsides and Downsides.' *Journal of Energy and Natural Resources Law* 26: 120–50.

Filer, C., D. Henton and R. Jackson, 2000. *Landowner Compensation in Papua New Guinea's Mining and Petroleum Sectors*. Port Moresby: PNG Chamber of Mines and Petroleum.

Filer, C. and R. Jackson, 1986. 'The Social and Economic Impact of a Gold Mine on Lihir.' Unpublished report to Lihir Liaison Committee.

——, 1989. *The Social and Economic Impact of a Gold Mine in Lihir: Revised and Expanded* (2 volumes). Konedobu: Lihir Liaison Committee.

Filer, C. and M. Macintyre, 2006. 'Grassroots and Deep Holes: Community Responses to Mining in Melanesia.' *Contemporary Pacific* 18: 215–32.

Filer, C. and A. Mandie- Filer, 1998. 'Rio Tinto Through the Looking Glass: Lihirian Perspectives on the Social and Environmental Aspects of the Lihir Gold Mine.' Unpublished report to World Bank.

Finney, B., 1973. *Big-Men and Business: Entrepreneurship and Economic Growth in the New Guinea Highlands*. Canberra: Australian National University Press.

Firth, S., 1976. 'The Transformation of the Labour Trade in German New Guinea, 1899–1914.' *Journal of Pacific History* 11: 51–65.

——, 1982. *New Guinea under the Germans*. Melbourne: Melbourne University Press.

Fitzpatrick, P., 1980. *Law and State in Papua New Guinea*. London: Academic Press.

Fortes, M. and E.E. Evans-Pritchard (eds), 1958 [1940]. *African Political Systems*. London: Oxford University Press.

Foster, R.J., 1993. 'Dangerous Circulation and Revelatory Display: Exchange Practices in a New Ireland Society.' In J. Fajans (ed.), *Exchanging Products: Producing Exchange*. Sydney: University of Sydney (Oceania Monograph 43).

——, 1995a. *Social Reproduction and History in Melanesia: Mortuary Ritual, Gift Exchange, and Custom in the Tanga Islands*. Cambridge: Cambridge University Press.

——, 1995b. 'Introduction.' In R.J. Foster (ed.), *Nation Making: Emergent Identities in Post-Colonial Melanesia*. Michigan: University of Michigan Press.

Foucault, M., 1975. *Discipline and Punish: The Birth of the Prison*. London: Penguin.

——, 1984. 'What is Enlightenment?' In P. Rabinow (ed.), *The Foucault Reader: An Introduction to Foucault's Thought*. London: Penguin Books.

Friedman, J., 1994. *Consumption and Identity*. Chur: Harwood Academic Publishers.

Gell, A., 1992. 'Inter-tribal Commodity Barter and Reproductive Gift-Exchange in Old Melanesia.' In C. Humphrey and S. Hugh-Jones (eds), *Barter, Exchange, and Value: An Anthropological Approach*. Cambridge: Cambridge University Press.

George, M. and D. Lewis, 1985. 'Maritime Trade and Traditional Exchange in the Bismarck Archipelago.' In J. Allen (ed.), *Lapita Homeland Project: Report of the 1985 Field Season*. Unpublished report.

Gerritsen, R. and M. Macintyre, 1986. *Social Impact Study of the Misima Gold Mine* (2 volumes). Boroko: Institute of Applied Social and Economic Research.

——, 1991. 'Dilemmas of Distribution: The Misima Gold Mine, Papua New Guinea.' In J. Connell and R. Howitt (eds), op. cit.

Gerritsen, R., R.J. May and M.A.H.B. Walter, 1982. *Road Belong Development: Cargo Cults, Community Groups and Self-Help Movements in Papua New Guinea*. Canberra: Australian National University, Research School of Pacific Studies, Department of Political and Social Change (Working Paper 3).

Geschiere, P., 1997. *The Modernity of Witchcraft: Politics and the Occult in Post-Colonial Africa*. Charlottesville: University Press of Virginia.

Gewertz, D. and F. Errington, 1995. 'Duelling Currencies in East New Britain: The Construction of Shell Money as National Cultural Property.' In J. Carrier (ed.), *Occidentalism: Images of the West*. Oxford: Clarendon Press.

——, 1999. *Emerging Class in Papua New Guinea: The Telling of Difference*. Cambridge: Cambridge University Press.

Gifford, P., 2004. *Ghana's New Christianity: Pentecostalism in a Globalising African Economy*. Bloomington: Indiana University Press.

Glaglas, L. and M. Soipang, 1993. 'Lihir Master Development Plan.' Unpublished typescript.

Glasse, R.M. and M.J. Meggitt, 1969. *Pigs, Pearlshells, and Women: Marriage in the New Guinea Highlands*. Englewood Cliffs (NJ): Prentice-Hall.

Godelier M., 1977. *Perspectives in Marxist Anthropology*. Cambridge: Cambridge University Press.

——, 1999. *The Enigma of the Gift*. Chicago: University of Chicago Press.

Godlier, M. and M. Strathern (eds). 1991. *Big Men and Great Men: Personifications of Power in Melanesia*. New York: Cambridge University Press.

Goldman, L., 2007. '"Hoo-Ha in Huli": Considerations on Commotion and Community in the Southern Highlands.' In N. Haley and R.J. May (eds), *Conflict and Resource Development in the Southern Highlands of Papua New Guinea*. Canberra: ANU E Press.

Golub, A., 2005. Making the Ipili Feasible: Imagining Local and Global Actors at the Porgera Gold Mine, Enga Province, Papua New Guinea. Chicago: University of Chicago (Ph.D. thesis).

Goodale, J., 1985. 'Pig's Teeth and Skull Cycles: Both Sides of the Face of Humanity.' *American Ethnologist* 12: 228–44.

Gouldner, A., 1960. 'The Norm of Reciprocity: A Preliminary Statement.' *American Sociological Review* 25: 161–78.

Gregory, C.A., 1982. *Gifts and Commodities*. London: Academic Press.

Groves, W.C., 1934. 'Tabar To-Day: A Study of a Melanesian Community in Contact with Alien Non-Primitive Cultural Forces.' *Oceania* 5: 224–40.

——, 1935. 'Tabar To-Day: A Study of a Melanesian Community in Contact with Alien Non-Primitive Cultural Forces.' *Oceania* 5: 346–60.

Gudeman, S., 1986. *Economics as Culture: Models and Metaphors of Livelihood*. London: Routledge & Kegan Paul.

Gunn, M., 1987. 'The Transfer of Malagan Ownership on Tabar.' In L. Lincoln (ed.), op. cit.

——, 1997. *Ritual Arts of Oceania: New Ireland in the Collections of the Barbier-Muller Museum*. Milan: Skira Editore.

Gunn, M. and P. Peltier, 2006. *New Ireland: Art of the South Pacific*. Milan: 5 Continents.

Handler, R. and J. Linnekin, 1984. 'Tradition, Genuine or Spurious?' *Journal of American Folklore* 97: 273–90.

Haro, B.V., 2010. The Impact of Personal Viability Training on Gender Relations in Mining Communities: The Case of Lihir, Papua New Guinea. Palmerston North (NZ): Massey University (Masters thesis).

Healey, C.J., 1985. 'New Guinea Inland Trade: Transformation and Resilience in the Context of Capitalist Penetration.' *Mankind* 15: 127–44.

Healy, A.M., 1967. *Bulolo: A History of the Development of the Bulolo Region, New Guinea*. Canberra and Port Moresby: Australian National University, Research School of Pacific Studies, New Guinea Research Unit (Bulletin 15).

Heaney, W., 1982. 'The Changing Role of Bird of Paradise Plumes in Bridewealth in the Wahgi Valley.' In L. Morauta, J. Pernetta and W. Heaney (eds), *Traditional Conservation in Papua New Guinea: Implications for Today*. Boroko: Institute of Applied Social and Economic Research (Monograph 16).

Hemer, S., 2001. A Malmalien e Makil: Person, Emotion and Relations in Lihir. Melbourne: University of Melbourne (Ph.D. thesis).

Herzfeld, M., 1990. 'Pride and Perjury: Time and the Oath in the Mountain Villages of Crete.' *Man* (N.S.) 25: 305–22.

Hogbin, I., 1944. 'Councils and Native Courts in the Solomon Islands.' *Oceania* 8: 257–83.

Horton, R., 1970. 'African Traditional Thought and Western Science.' In B. Wilson (ed.), *Rationality*. Oxford: Basil Blackwell.

Howitt, R., J. Connell and P. Hirsch (eds), 1996. *Resources, Nations and Indigenous Peoples: Case Studies from Australasia, Melanesia and Southeast Asia*. Melbourne: Oxford University Press.

Hughes, I., 1978. 'Good Money and Bad: Inflation and Devaluation in the Colonial Process.' *Mankind* 11: 308–18.

ICMM (International Council for Mining and Metals), 2003. 'Sustainable Development Framework.' Viewed 24 June 2010 at www.icmm.com

IFC (International Finance Corporation), 2006. 'IFC Environmental and Social Performance Standards.' Viewed 24 June 2010 at www.ifc.org

Imbun, B.Y., 2000. 'Mining Workers or "Opportunist" Tribesmen? A Tribal Workforce in a Papua New Guinea Mine.' *Oceania* 71: 129–49.

Jacka, J., 2001. 'On the Outside Looking In: Attitudes and Responses of Non-Landowners towards Mining at Porgera.' In B.Y. Imbun and P.A. McGavin (eds), *Mining in Papua New Guinea: Analysis & Policy Implications*. Waigani: University of Papua New Guinea Press.

Jackson, R., 1997. 'Cheques and Balances: Compensation and Mining in Papua New Guinea.' In S. Toft (ed), op. cit.

——, 2000. 'Kekeisi Kekeisi: A Long Term Economic Development Plan for the Misima Gold Mine's Impact Area.' Unpublished consultancy report.

Jackson, R. and G. Banks, 2002. *In Search of the Serpent's Skin: A History of the Porgera Gold Mine*. Port Moresby: Placer Niugini Ltd.

Jessep, O.D., 1977. Land Tenure in a New Ireland Village. Canberra: Australian National University (Ph.D. thesis).

Jolly, M., 1992. 'Spectres of Inauthenticity.' *Contemporary Pacific* 4: 49–72.

——, 1994. *Women of the Place: Kastom, Colonialism, and Gender in Vanuatu*. Philadelphia: Harwood Academic Publishers.

——, 1997. 'Woman-Nation-State in Vanuatu: Women as Signs and Subjects in the Discourses of Kastom, Modernity and Christianity.' In T. Otto and N. Thomas (eds), *Narratives of the Nation in the South Pacific*. Amsterdam: Harwood Academic Publishers.

Jorgensen, D., 1981. 'Life on the Fringe: History and Society in Telefolmin.' In R. Gordon (ed.), *The Plight of Peripheral People in Papua New Guinea* ⊠*Volume I: The Inland Situation*. Cambridge (MA): Cultural Survival.

———, 1997. 'Who and What Is a Landowner? Mythology and Marking the Ground in a Papua New Guinea Mining Project.' *Anthropological Forum* 7: 599–627.

Kabariu, L., n.d. 'The Alaia Sacred Site.' Unpublished typescript.

Kahn, M., 1983. 'Sunday Christians, Monday Sorcerers: Selective Adaptation to Missionization in Wamira.' *Journal of Pacific History* 18: 96–122.

Kaplan, S., 1976. 'Ethnological and Biogeographical Significance of Pottery Shards from Nissan Island, Papua New Guinea.' *Fieldiana - Anthropology* 66: 35–89.

Keesing, R., 1976. *Cultural Anthropology*. New York: Holt, Rinehart and Winston.

———, 1982. 'Kastom in Melanesia: An Overview.' *Mankind* 13: 297–301.

———, 1989. 'Creating the Past: Custom and Identity in the Contemporary Pacific.' *Contemporary Pacific* 1: 19–42.

Kirsch, S., 2004. 'Keeping the Network in View: Compensation Claims, Property and Social Relations in Melanesia.' In L. Kalinoe and J. Leach (eds), *Rationales of Ownership: Transactions and Claims to Ownership in Contemporary Papua New Guinea*. Wantage: Sean Kingston Publishing.

———, 2006. *Reverse Anthropology: Indigenous Analysis of Social and Environmental Relations in New Guinea*. Stanford: Stanford University Press.

Knauft, B.M., 1999. *From Primitive to Postcolonial in Melanesia and Anthropology*. Michigan: University of Michigan Press.

———, 2002. 'Critically Modern: An Introduction.' In B.M. Knauft (ed.), *Critically Modern: Alternatives, Alterities, Anthropologies*. Bloomington: Indiana University Press.

Koczberski, G. and G. Curry, 2004. 'Divided Communities and Contested Landscapes: Mobility, Development and Shifting Identities in Migrant Sites in Papua New Guinea.' *Asia Pacific Viewpoint* 45: 357–71.

Kopytoff, I., 1986. 'The Cultural Biography of Things: Commoditization as Process.' In A. Appadurai (ed), *The Social Life of Things: Commodities in Cultural Perspective*. Cambridge: Cambridge University Press.

Koselleck, R., 1985. *Futures Past: On the Semantics of Historical Time*. Cambridge (MA): MIT University Press.

Kowal, E., 1999. Children, Choice and Change: The Use of Western Contraception on the Lihir Islands, Papua New Guinea. Melbourne: University of Melbourne (Masters thesis).

Kramer-Bannow, E., 2008 [1916]. *Among Art-Loving Cannibals of the South Seas*. Adelaide: Crawford House.

Kroeber, A.L., 1938. 'Basic and Secondary Patterns of Social Structure.' *Journal of the Royal Anthropological Institute* 45: 299–310.

Küchler, S., 2002. *Malanggan: Art, Memory and Sacrifice*. Oxford: Berg Publishers.

Lagisa, L., 1997. Major Development Project on Women's Lives: A Case Study of Mining in Lihir, Papua New Guinea. Palmerston North (NZ): Massey University (Masters thesis).

Lattas, A., 1998. *Cultures of Secrecy: Reinventing Race in Bush Kaliai Cargo Cults*. Madison: University of Wisconsin Press.

——, 2007. 'Cargo Cults and the Politics of Alterity: A Review Article.' *Anthropological Forum* 17: 149–61.

Lawrence, P., 1964. *Road Belong Cargo: A Study of the Cargo Movement in the Southern Madang District New Guinea*. Melbourne: Melbourne University Press.

Lawrence, P. and M.L. Meggitt (eds), 1965. *Gods Ghosts and Men in Melanesia: Some Religions of Australian New Guinea and the New Hebrides*. Melbourne: Oxford University Press.

Leach, E.R. and J.W Leach (eds), 1983. *The Kula: New Perspectives on Massim Exchange*. Cambridge: Cambridge University Press.

Leavitt, S.C., 2000. 'The Apotheosis of White Men?: A Re-Examination of Beliefs about Europeans as Ancestral Spirits.' *Oceania* 70: 304–24.

Levi-Strauss, C., 1969 [1949]. *The Elementary Structures of Kinship*. Boston: Beacon.

Lincoln, L., 1987. *Assemblage of Spirits: Idea and Image in New Ireland*. New York: George Braziller.

Lindstrom, L., 1984. 'Doctor, Lawyer, Wise Man, Priest: Big Men and Knowledge in Melanesia.' *Man* (N.S.) 19: 291–309.

——, 1990. *Knowledge and Power in a South Pacific Society*. Washington: Smithsonian Press.

——, 1995. 'Cargoism and Occidentalism.' In J. Carrier (ed.), *Occidentalism: Images of the West*. Oxford: Clarendon Press.

LiPuma, E., 2000. *Encompassing Others: the Magic of Modernity in Melanesia*. Michigan: University of Michigan Press.

LJNC (Lihir Joint Negotiating Committee). 2004a. 'Stakeholders Presentation on the Road Map for the Completion of the New IBP.' Unpublished typescript.

——, 2004b. 'A Peketon ⊠ "Be Viable": Master Plan for Lihir Grasruts Pawa Mekim Kamap Limited, incorporating Lihir Sustainable Development Plan.' Unpublished typescript.

LMALA (Lihir Mining Area Landowners Association), 1994. 'Lihir Society Reform Implementation Programme.' Unpublished typescript.

——, n.d. 'The Definition of a Lihirian.' Unpublished typescript.

Lockwood, V.S., 2004. 'The Global Imperative and Pacific Island Societies.' In V.S. Lockwood (ed.), *Globalization and Culture Change in the Pacific Islands*. Upper Saddle River (NJ): Pearson Prentice Hall.

LSDP (Lihir Sustainable Development Plan), 2007. 'Executive Summary: Lihir Sustainable Development Plan.' Unpublished typescript.

LWCC (Lihir Working Cultural Committee), 2004. 'Major and Minor Customs and Customary Laws.' Unpublished typescript.

Macintyre, M., 1983. Changing Paths: An Historical Ethnography of Traders of Tubetube. Canberra: Australian National University (Ph.D. thesis).

——, 1984. 'The Problem of the Semi-Alienable Pig.' *Canberra Anthropology* 7: 109–21.

——, 1996. 'Lihir Gold Mine Project: Social, Political and Economic Impact.' Unpublished report to Lihir Gold Limited.

——, 1997. 'Social and Economic Impact Study and Monitoring, Lihir, New Ireland Province.' Unpublished report to Lihir Gold Limited.

——, 1998a. 'Social and Economic Impact Study, Lihir 1998.' Unpublished report to Lihir Gold Limited.

——, 1998b. 'The Persistence of Inequality: Women in Papua New Guinea since Independence.' In L. Zimmer-Tamakoshi (ed.), op. cit.

——, 1999. 'Social and Economic Impact Study Lihir 1999.' Unpublished report to Lihir Gold Limited.

——, 2000. 'Hear Us, Women of Papua New Guinea!: Melanesian Women and Human Rights.' In A.M. Hildson, M. Macintyre, V. Mackie and M. Stivens (eds), *Human Rights and Gender Politics: Asia-Pacific Perspectives*. New York: Routledge.

——, 2003. 'Peztorme Women: Responding to Change in Lihir, Papua New Guinea.' *Oceania* 74: 120–33.

——, 2006. 'Women Working in the Mining Industry in Papua New Guinea: A Case Study from Lihir.' In K. Lahiri-Dutt and M. Macintyre (eds), *Women Miners in Developing Countries: Pit Women and Others*. Aldershot: Ashgate Publishing.

——, 2008. 'Police and Thieves, Gunmen and Drunks: Problems with Men and Problems with Society in Papua New Guinea.' *Australian Journal of Anthropology* 19: 179–93.

Macintyre, M. and S. Foale, 2000. 'Social and Economic Impact Study Lihir 2000.' Unpublished report to Lihir Gold Limited.

——, 2001. 'Social and Economic Impact Study Lihir 2001.' Unpublished report to Lihir Gold Limited.

——, 2003. 'Social and Economic Impact Study Lihir 2003.' Unpublished report to Lihir Gold Limited.

——, 2004. 'Global Imperatives and Local Desires: Competing Economic and Environmental Interests in Melanesian Communities.' In V.S. Lockwood (ed.) *Globalization and Culture Change in the Pacific Islands*. Upper Saddle River (NJ): Pearson Prentice Hall.

——, 2007. 'Land and Marine Tenure, Ownership and New Forms of Entitlement on Lihir: Changing Notions of Property in the Context of a Gold Mining Project.' *Human Organization* 66: 49–59.

Macpherson, C.B., 1962. *The Political Theory of Possessive Individualism: Hobbes to Locke*. Oxford: Clarendon Press.

Malinowski, B., 1961 [1922]. *Argonauts of the Western Pacific: An Account of Native Enterprise and Adventure in the Archipelagoes of Melanesian New Guinea*. London: Routledge & Kegan Paul.

Marshall, M., 1979. *Weekend Warriors: Alcohol in a Melanesian Culture*. Palo Alto: Mayfield Publishing Company.

—— (ed.), 1982. *Through a Glass Darkly: Beer and Modernization in Papua New Guinea*. Boroko: Institute of Applied Social and Economic Research (Monograph 18).

Marx, K., 1909. *Capital: A Critique of Political Economy ⊠ Volume 3: The Process of Capitalist Production as a Whole*. Chicago: Charles H. Kerr & Company.

Mauss, M., 2002 [1925]. *The Gift: The Form and Reason for Exchange in Archaic Societies*. London: Routledge.

May, R.J. (ed.), 1982. *Micronationalist Movements in Papua New Guinea*. Canberra: Australian National University Research School of Pacific Studies, Department of Political and Social Change (Monograph 1).

——, 2001. *State and Society in Papua New Guinea: The First Twenty-Five Years*. Canberra: ANU E Press.

May, R.J. and M. Spriggs, 1990. *The Bougainville Crisis*. Bathurst: Crawford House.

MCA (Minerals Council of Australia), 2006. 'Unearthing New Resources: Attracting and Retaining Women in the Australian Minerals Industry.' Canberra: MCA.

Mead, M., 1956. *New Lives for Old: Cultural Transformation — Manus, 1928–1953*. New York: William Morrow and Company.

Meggitt, M.J., 1973. '"Pigs are Our Hearts!" The Te Exchange Cycle among the Mae Enga of New Guinea.' *Oceania* 44: 165–203.

Meillassoux, C., 1981. *Maidens, Meal, and Money: Capitalism and the Domestic Community*. New York: Cambridge University Press.

Meyer, B., 1998. 'Make a Complete Break with the Past: Memory and Post-Colonial Modernity in Ghanaian Pentecostalist Discourse.' *Journal of Religion in Africa* 28: 316–49.

——, 1999. *Translating the Devil: Religion and Modernity among the Ewe in Ghana*. Edinburgh: Edinburgh University Press.

——, 2004. 'Christianity in Africa: From African Independent to Penticostal-Charismatic Churches.' *Annual Review of Anthropology* 33: 447–74.

Miller, D., 1987. *Material Culture and Mass Consumption*. Oxford: Blackwell.

Miskaram, N., 1985. 'Cargo Cultism on New Hanover: A Psychopathological Phenomenon or an Indication of Unequal Development?' In C. Loeliger and

G. Trompf (eds), *New Religious Movements in Melanesia*. Suva: University of the South Pacific , Institute of Pacific Studies. Port Moresby: University of Papua New Guinea.

MML (Misima Mines Limited), 2000. 'Misima Mines Limited Sustainability Plan 2000.' Port Moresby: MML.

Modjeska, N., 1982. 'Production and Inequality: Perspectives from Central New Guinea.' In A. Strathern (ed.), *Inequality in New Guinea Highlands Societies*. Cambridge: Cambridge University Press.

Munn, N., 1986. *The Fame of Gawa: A Symbolic Study of Value Transformation in a Massim (Papua New Guinea) Society*. Cambridge: Cambridge University Press.

Nalu, M., 2006. 'Grasruts Universiti.' *Paradise* 1: 46–7.

Narokobi, B., 1980. *The Melanesian Way*. Boroko: Institute of Papua New Guinea Studies. Suva: University of the South Pacific, Institute of Pacific Studies.

Nash, J., 1987. 'Gender Attributes and Equality: Men's Strength and Women's Talk among the Nagovisi.' In M. Strathern (ed.), *Dealing with Inequality: Analysing Gender in Melanesia and Beyond*. Cambridge: Cambridge University Press.

Nash, J. and E. Ogan, 1990. 'The Red and the Black: Bougainvillean Perceptions of Other Papua New Guineans.' *Pacific Studies* 13: 1–17.

Nelson, H., 1976. *Black, White and Gold: Goldmining in Papua New Guinea 1878–1930*. Canberra: Australian National University Press.

Newbury, C., 1975. 'Colour Bar and Labour Conflict on the New Guinea Goldfields 1935–41.' *Australian Journal of Politics and History* 21: 25–38.

NRLLG (Nimamar Rural Local Level Government), 2002. 'The Sengseng (Movement) Monitoring Policy Action Plan.' Unpublished typescript.

Obeyesekere, G., 2005. *Cannibal Talk: The Man-Eating Myth and Human Sacrifice in the South Seas*. Berkeley: University of California Press.

Oliver, D.L., 1967 [1955]. *A Solomon Islands Society: Kinship and Leadership among the Siuai of Bougainville*. Boston: Beacon Press.

Parkinson, R., 1999 [1907]. *Thirty Years in the South Seas: Land and People, Customs and Traditions in the Bismarck Archipelago and on the German Solomon Islands*. Bathurst: Crawford House.

Polanyi, K., 2001 [1944]. *The Great Transformation: The Political and Economic Origins of Our Time*. Boston: Beacon Press.

Powdermaker, H., 1971 [1933]. *Life in Lesu: The Study of a Melanesian Society in New Ireland*. New York: W.W. Norton & Co.

Price, C.A. and E. Baker, 1976. 'Origins of Pacific Island Labourers in Queensland, 1863–1906: A Research Note.' *Journal of Pacific History* 11: 106–21.

Ramstad, Y., n.d 1. 'The TKA Movement in New Ireland.' Paper presented to the Department of Anthropology and Sociology, Research School of Pacific Studies, Australian National University.

——, n.d. 2. 'Introducing Lihir Society: Kinship Roles and Human Resources.' Paper presented to the Department of Anthropology and Sociology, Research School of Pacific Studies, Australian National University.

——, n.d. 3. 'Rituals on Lihir.' Paper presented to the Department of Anthropology and Sociology, Research School of Pacific Studies, Australian National University.

Rappaport, R.A., 1968. *Pigs for the Ancestors: Ritual in the Ecology of a New Guinea People*. London: Yale University Press.

Robbins, J., 2004. *Becoming Sinners: Christianity and Moral Torment in a Papua New Guinean Society*. Berkeley: University of California Press

——, 2005. 'Humiliation and Transformation: Marshall Sahlins and the Study of Cultural Change in Melanesia.' In J. Robbins and H. Wardlow (eds), *The Making of Global and Local Modernities in Melanesia: Humiliation, Transformation and the Nature of Cultural Change*. Aldershot: Ashgate Publishing.

Rowley, C.D., 1958. *The Australians in German New Guinea, 1914–1921*. Melbourne: Melbourne University Press.

Ryan, P., 1991. *Black Bonanza: A Landslide of Gold*. South Yarra: Hyland House.

Sack, P. and D. Clark (eds), 1978. *German New Guinea: The Annual Reports*. Canberra: Australian National University Press.

Sahlins, M., 1963. 'Poor Man, Rich Man, Big-Man, Chief: Political Types in Melanesia and Polynesia.' *Comparative Studies in Society and History* 5: 285–303.

——, 1965. 'On the Sociology of Primitive Exchange.' In M. Banton (ed.), *The Relevance of Models for Social Anthropology*. London: Tavistock Publications.

———, 1974. *Stone Age Economics*. London: Tavistock Publications.

———, 1985. *Islands of History*. Chicago: University of Chicago Press.

———, 1992. 'The Economics of Develop-Man in the Pacific.' *Res* 21: 13–25.

———, 1999. 'Two or Three Things I Know about Culture.' *Journal of the Royal Anthropological Institute* (N.S.) 5: 399–421.

———, 2004. *Apologies to Thucydides: Understanding History as Culture and Vice Versa*. Chicago: University of Chicago Press.

———, 2005a. 'Introduction.' In M. Sahlins (ed.), *Culture in Practice: Selected Essays*. New York: Zone Books.

———, 2005b. 'Cosmologies of Capitalism: The Trans-Pacific Sector of the "The World System".' In M. Sahlins (ed.), *Culture in Practice: Selected Essays*. New York: Zone Books.

———, 2005c. 'Goodbye to Tristes Tropes: Ethnography in the Context of Modern World History.' In M. Sahlins (ed.), *Culture in Practice: Selected Essays*. New York: Zone Books.

Salisbury, R.F., 1962. *From Stone to Steel: Economic Consequences of a Technological Change in New Guinea*. Melbourne: Melbourne University Press.

———, 1970. *Vunamami: Economic Transformation in a Traditional Society*. Melbourne: Melbourne University Press.

Scarr, D., 1967. *Fragments of Empire: A History of the Western Pacific High Commission 1887–1914*. Canberra: Australia National University Press.

Schlaginhaufen, O., 1959. *Muliama: Zwei Jahre unter Sudsee Insulanern*. Zurich: Orell Fussli Verlag.

Schieffelin, E.L. and D. Gewertz, 1985. 'Introduction.' In D. Gewertz and E. Schieffelin (eds), *History and Ethnohistory in Papua New Guinea*. Sydney: University of Sydney (Oceania Monograph 28).

Schwartz, T., 1962. 'The Paliau Movement in the Admiralty Islands, 1946–1954.' *Anthropological Papers of the American Museum of Natural History* 49: 207–421.

———, 1963. 'Systems of Areal Integration.' *Anthropological Forum* 1: 56–97.

Seligman, C., 1910. *The Melanesians of British New Guinea*. Cambridge: Cambridge University Press.

Sharp, A., 1968. *The Voyages of Abel Janszoon Tasman*. Oxford: Clarendon Press.

Siegel, J., 1985. 'Origins of Pacific Islands Labourers in Fiji.' *Journal of Pacific History* 20: 42–54.

Simmel, G., 1978 [1900]. *The Philosophy of Money*. London: Routledge & Kegan Paul.

Skalnik, P., 1989. 'Lihir Society on the Eve of Gold Mine Operations: A Proposal for Urgent Anthropological Research.' Unpublished typescript.

Smalley, I., 1985. 'Sociological Baseline Study of Lihir Island, New Ireland Province, PNG.' Unpublished report to Kennecott-Niugini Mining Joint Venture.

Strathern, A., 1971a. 'Cargo and Inflation in Mount Hagen.' *Oceania* 41: 255–65.

———, 1971b. *The Rope of Moka: Big-Men and Ceremonial Exchange in Mount Hagen, New Guinea*. Cambridge: Cambridge University Press.

———, 1979. 'Gender, Ideology, and Money in Mount Hagen.' *Man* (N.S.) 14: 530–48.

———, 1982a. 'Alienating the Inalienable.' *Man* (N.S.) 17: 548–51.

———, 1982b. 'The Division of Labour and Processes of Social Change in Mount Hagen.' *American Ethnologist* 9: 307–19.

Strathern, M., 1985. 'Knowing Power and Being Equivocal: Three Melanesian Contexts. In R. Fardon (ed.), *Power and Knowledge: Anthropological and Sociological Approaches*. Edinburgh: Scottish Academic Press.

Tam, S., 1997. 'EDTC Personal Viability: Holistic Human Development ☒ Physical-Mental-Spiritual-Emotional-Financial.' Boroko: Entrepreneurial Development Training Centre Ltd.

———, 2010. Entrepreneurial Development Training Centre Website. Viewed 24 June 2010 at http://edtc.ac.pg

Taussig, M., 1980. *The Devil and Commodity Fetishism in South America*. Chapel Hill: University of North Carolina Press.

Thomas, N., 1991. *Entangled Objects: Exchange, Material Culture, and Colonialism in the Pacific*. Cambridge (MA): Harvard University Press.

Toft, S. (ed.), 1997. *Compensation for Resource Development in Papua New Guinea.* Port Moresby: Law Reform Commission of Papua New Guinea (Monograph 6). Canberra: Australian National University, National Centre for Development Studies (Pacific Policy Paper 24).

Trompf, G., 1991. *Melanesian Religion.* Cambridge: Cambridge University Press.

Trouillot, M-R., 2002. 'The Otherwise Modern: Caribbean Lessons from the Savage Slot.' In B. Knauft (ed.), *Critically Modern: Alternatives, Alterities, Anthropologies.* Bloomington: Indiana University Press.

Unage, M., 2006. 'The Way to Human Resource Development.' *The National,* 26 April.

UN (United Nations), 2000. 'Corporate Citizenship in the World Economy: The Global Compact.' Viewed 24 June 2010 at www.unglobalcompact.org

Vail, J., 1995. 'All That Glitters: The Mt Kare Gold Rush and Its Aftermath.' In A. Biersack (ed.), *Papuan Borderlands: Huli, Duna, and Ipili Perspectives on the Papua New Guinea Highlands.* Ann Arbor: University of Michigan Press.

Van Helden, F. 1998. *Between Cash and Conviction: The Social Context of the Bismarck-Ramu Integrated Conservation and Development Project.* Boroko: National Research Institute (Monograph 33).

Wagner, R., 1975. *The Invention of Culture.* Chicago: University of Chicago Press.

——, 1986. *Asiwinarong: Ethos, Image, and Social Power among the Usen Barok of New Ireland.* Princeton: Princeton University Press.

Wallerstein, I., 1974. *The Modern World-System: Capitalist Agriculture and the Origins of the European World-Economy in the Sixteenth Century.* New York: Academic Press.

Walter, M., 1981. 'Cult Movements and Community Development Associations: Revolution and Evolution in the Papua New Guinea Countryside.' In R. Gerritsen, R.J. May and M. Walter (eds), op. cit.

Ward, G. and E. Kingdon, 1995. 'Land Tenure in the Pacific Islands.' In G. Ward and E. Kingdon (eds), *Land, Custom and Practice in the South Pacific.* Cambridge: Cambridge University Press.

Wardlow, H., 2002. '"Hands-Up"-ing Buses and Harvesting Cheese-Pops: Gendered Mediation of Modern Disjuncture in Melanesia.' In B. Knauft (ed.), *Critically Modern: Alternatives, Alterities, Anthropologies.* Bloomington: Indiana University Press.

——, 2006. *Wayward Women: Sexuality and Agency in a New Guinea Society.* Berkeley: University of California Press.

Weber, M., 1978. *Economy and Society: An Outline of Interpretative Sociology* (2 volumes) (ed. G. Roth and C. Wittich). Berkeley: University of California Press.

Weiner, A., 1976. *Women of Value, Men of Renown.* Brisbane: University of Queensland Press.

Weiner, J.F., 1991. *The Empty Place: Poetry, Space, and Being among the Foi of Papua New Guinea.* Bloomington: Indiana University Press.

Weiner, J.F. and K. Glaskin (eds), 2007. *Customary Land Tenure and Registration in Australia and Papua New Guinea: Anthropological Perspectives.* Canberra: ANU E Press.

West, P., 2006. *Conservation Is Our Government Now: The Politics of Ecology in Papua New Guinea.* Durham (NC): Duke University Press.

Williams, F.E., 1979 [1923]. 'The Vailala Madness and the Destruction of Native Ceremonies in the Gulf Division.' In F.E. Williams, *"The Vailala Madness" and Other Essays* (ed. E. Schwimmer). Honolulu: University of Hawai'i Press.

Wolf, E., 1982. *Europe and the People without History.* Berkeley: University of California Press.

Wolfers, E.P., 1992. 'Politics, Development and Resources: Reflections on Constructs, Conflicts and Consultants.' In S. Henningham and R.J. May (eds), *Resources, Development and Politics in the Pacific Islands.* Bathurst: Crawford House.

Worsley, P., 1968. *The Trumpet Shall Sound: A Study of 'Cargo' Cults in Melanesia.* New York: Schocken Books.

Young, M., 1971. *Fighting with Food: Leadership, Values and Social Control in a Massim Society.* Cambridge: Cambridge University Press.

Zial, S., 1975. 'Lihir Experiences under the Japanese.' *Oral History* 3: 66–76.

Zimmer-Tamakoshi, L. (ed.), 1998. *Modern Papua New Guinea.* Kirksville (MO): Thomas Jefferson University Press.

www.ingramcontent.com/pod-product-compliance
Lightning Source LLC
Chambersburg PA
CBHW061244270326
41928CB00041B/3408